In Vitro Bioassay Techniques for Anticancer Drug Discovery and Development

In Vitro Bioassay Techniques for Anticancer Drug Discovery and Development

Dhanya Sunil, Pooja R. Kamath, and
H. Raghu Chandrashekhar

CRC Press
Taylor & Francis Group
Boca Raton   London   New York

CRC Press is an imprint of the
Taylor & Francis Group, an **informa** business

CRC Press
Taylor & Francis Group
6000 Broken Sound Parkway NW, Suite 300
Boca Raton, FL 33487-2742

First issued in paperback 2022

ISBN-13: 978-1-138-72582-9 (hbk)
ISBN-13: 978-1-03-233968-9 (pbk)
DOI: 10.4324/9781315191256

Library of Congress Cataloging-in-Publication Data

Names: Sunil, Dhanya, author. | Kamath, Pooja R., author. | Chandrashekhar, H. Raghu, author.
Title: In vitro bioassay techniques for anticancer drug discovery and development / Dhanya Sunil, Pooja R. Kamath and H. Raghu Chandrashekhar
Description: Boca Raton : CRC Press, [2017] | Includes bibliographical references and index.
Identifiers: LCCN 2016055787| ISBN 9781138725829 (hardback : alk. paper) | ISBN 9781315191256 (ebook)
Subjects: LCSH: Antineoplastic agents--Research--Methodology. | Electrochemical sensors.
Classification: LCC RC271.C5 S86 2017 | DDC 616.99/4061072--dc23
LC record available at https://lccn.loc.gov/2016055787

Visit the Taylor & Francis Web site at
http://www.taylorandfrancis.com

and the CRC Press Web site at
http://www.crcpress.com

Contents

Preface

The process of anticancer drug development demands a multidisciplinary collaboration between biologists, medicinal chemists, pharmacologists, toxicologists, and clinicians. Simple and rapid bioassays can serve as starting points for such multidisciplinary efforts which can then be followed by more detailed mechanism-based studies targeted at anticancer drug discovery. The book offers the reader an insight into the intricate processes in drug discovery and the importance and role of *in vitro* bioassays for screening compounds and their specific applications in the assessment of therapies for cancer.

This handbook represents a definitive, up-to-date compendium of the key *in vitro* bioassay methods that are employed to quantify and validate the anticancer activity of a drug candidate before it makes its way into animal or clinical trials. The book covers the screening and evaluation of potential drug candidates in a wide category of anticancer assays demonstrating the specific ways in which various pharmaceutical bioassays interpret the activity of drug molecules. The major emphasis of the book is to present those bioassays which can be readily set up and practiced in any laboratory with limited funds, facilities, or technical know-how.

An anticancer drug can hinder cancer dissemination via different mechanisms— by targeting the cell cycle, preventing malignant cell proliferation, hampering metastasis by obstructing cancer cell migration and invasion, or by inhibiting angiogenesis required for cancer cell nourishment. This book is an easy-to-follow guide which encompasses nine chapters. The first chapter introduces the reader to the various stages and the complexities involved in bringing a potential anticancer drug candidate into the market. The second chapter elaborates on the relevance of *in vitro* bioassays in anticancer drug development, highlighting its major strengths and limitations compared to *in vivo* methods. It also focuses on basic cell culture techniques and the necessary precautions that need to be taken during the process. The third chapter explains the significant role of cell viability and cytotoxicity assays, the principles involved, along with the protocols, advantages, and limitations. The fourth chapter details the need for cell proliferation assays and the various procedures employed in assessing the effect of drug candidates on the proliferation of cells. The fifth chapter deals with the different types of cell death and the features that differentiate one from the other. The bioassays that help in confirming the apoptotic mode of cell death are discussed in detail. The sixth and the seventh chapters expand on the assays that help to identify the antimigratory and anti-invasive capabilities of a test compound. The eighth chapter focuses on the assays that reveal the anti-angiogenic potential of a drug candidate. The ninth and final chapter renders an overview on the high-throughput anticancer drug discovery process. The illustrations of each assay together with the schematic diagrams presented in a simple manner in this book will aid the reader in comprehending the significance of the assays and hence to choose the most appropriate one or a combination of assays to screen the molecules and to study the mechanism for anticancer activity.

This book is a reference manual for any pharmaceutical researcher or medicinal chemist as it highlights the principles behind clinically relevant methods, ideas, and techniques belonging to a specific category. It will also guide them to carry out these assays employing the protocols given and to appropriately choose the tests according to the parameters that need to be assessed in the study.

Dhanya Sunil
Pooja R. Kamath
H. Raghu Chandrashekhar

1 Introduction to Anticancer Drug Development

1.1 CANCER

Cancer is a genomic disease with complex mechanisms which is hard to be understood, where the cells in a particular tissue change uncharacteristically, become less specialized, and do not respond to the signals generated inside the tissue that controls cellular differentiation, proliferation, survival, and finally cell death [1,2]. Cancer cells are capable to evading immune systems and spreading to distant areas or invading surrounding tissues [3]. There are more than 100 different types of cancer that are reported to date, and each one has its own methods of diagnosis and treatment [4].

The incidence of cancer is found to increase globally, attributed mainly to aging and genetic factors. There is also a contribution due to DNA damage caused by a cancer-associated lifestyle, which includes smoking, tobacco consumption, physical inactivity, certain dietary habits, exposure to radiation, and so on [5]. Cancer is the principal cause of death reported in economically developed countries and the second reason for death in developing countries [6].

1.2 CHEMOTHERAPY

Treatment of cancer is decided primarily depending on the type of cancer, site of its origin, or whether it is metastasized to other parts of the body or the stage of cancer the patient is suffering from [7]. Cancer is mainly managed through any one of the following treatment modalities like surgery, immunotherapy, chemotherapy, radiation therapy, hormone therapy, targeted therapy, or their combinations [8]. Chemotherapy, the most common treatment strategy, uses drugs to either stop or slow down the growth of cancer cells or kill them [9].

Though there are many side effects from chemotherapy, it is the most preferred treatment strategy because usually cancer in patients is diagnosed once it starts spreading to the entire body during which other cancer therapies will not help in controlling cancer dissemination [10]. As chemotherapy affects cell division, and a larger percentage of the targeted cells are undergoing cell division at any point in time, cells having higher growth rates are more affected by chemotherapy, whereas malignant tumors with lower growth rates tend to be less sensitive to chemotherapeutic agents. Heterogenic tumors, depending on the tumor populations, may respond to chemotherapy drugs displaying varying sensitivities [11].

1.3   NEED FOR EFFECTIVE ANTICANCER DRUGS

Existing marketed anticancer chemotherapeutic agents suffer from a broad range of issues associated with their efficacy and safety profile [12]. Cytotoxic chemotherapeutics primarily affect the cells that are rapidly proliferating, mostly sparing the cancer cells that are in the resting phase. Moreover, most of these cytotoxic anticancer agents only influence a cell's ability to multiply, with little effect on other features of tumor development such as progressive loss of cellular differentiation, tissue invasion, or finally metastases, which are considered the hallmark of malignant tumors. The predominant reason for the clinical death of most cancer patients is attributed to invasion and metastasis that involve multistep biochemical processes like cell detachment from the primary tumor site, invasion, migration to other body areas via intravasation and circulation, and implantation as secondary tumors followed by angiogenesis and proliferation [13,14]. Drugs that inhibit the spread of cancer to secondary sites by affecting any of the above features of metastatic dissemination will be broadly useful [15,16]. Also, the cytotoxic agents are associated with a high incidence of important adverse effects, which include alopecia, bone marrow suppression, mucositis, nausea, and vomiting. Because of all these factors, there is a high therapeutic urge to find newer anticancer agents that can focus individually or collectively on different characteristic features of malignant cells from different possible ways, to improve the outcome of cancer chemotherapy and thereby combat cancer.

1.4   STAGES OF ANTICANCER DRUG DEVELOPMENT

The increase in the techniques employed in the discovery of new anticancer therapies and investigations can be ascribed to the improved conception of cancer biology mechanisms. The various anticancer therapies and chemotherapeutic agents interact with their respective target by blocking the vital pathways involved in cancer cell proliferation or metastasis or activate cell death pathways. Every new anticancer drug developed, whether a single drug, hybrid drug, or a drug combination, is subjected to safety and efficacy assessment before being approved by the regulatory bodies [17].

As per the National Cancer Institute's (NCI) drug development program, the steps involved in developing an anticancer chemotherapeutic are as follows:

Preliminary in vitro *screening*: New drug candidates are first prescreened in a panel of three human tumor cell lines (breast MX-1, lung LX-1, and colon CX-1) at a single concentration for 48 h. If the drug inhibits the growth of any one or more cell lines, then its inhibition efficacy is further tested against a comprehensive panel of 60 human tumor cell lines [18].

In vitro *screening in human tumors*: This *in vitro* drug discovery selection method is proposed to replace transplantable animal tumors employed in anticancer drug screening process by 60 human tumor cell line panels that include lung, kidney, prostate, colon, breast, ovarian, and melanoma cancers as presented in Table 1.1. These cell lines sufficiently meet the

TABLE 1.1
NCI-60 Cell Lines for Anticancer Drug Screening

Panel Name	Cell Line Name	Doubling Time (h)	Panel Name	Cell Line Name	Doubling Time (h)
Leukemia	CCRF-CEM	26.7	CNS	SF-268	33.1
	HL-60(TB)	28.6		SF-295	29.5
	K-562	19.6		SF-539	35.4
	MOLT-4	27.9		SNB-19	34.6
	RPMI-8226	33.5		SNB-75	62.8
	SR	28.7		U251	23.8
Non-small	A549	22.9	Melanoma	LOX IMVI	20.5
cell lung	EKVX	43.6		MALME-3M	46.2
	HOP-62	39		M14	26.3
	NCI-H226	61		MDA-MB-435	25.8
	NCI-H23	33.4		SK-MEL-2	45.5
	NCI-H322M	35.3		SK-MEL-28	35.1
	NCI-H460	17.8		SK-MEL-5	25.2
	NCI-H522	38.2		UACC-257	38.5
Colon	COLO 205	23.8		UACC-62	31.3
	HCC-2998	31.5	Ovarian	IGR-OV1	31
	HCT-116	17.4		OVCAR-3	34.7
	HCT-15	20.6		OVCAR-4	41.4
	HT29	19.5		OVCAR 5	48.8
	KM12	23.7		OVCAR-8	26.1
	SW-620	20.4		NCI/ADR-RES	34
Prostate	PC-3	27.1		SK-OV-3	48.7
	DU-145	32.3	Renal	786-0	22.4
Breast	MCF7	25.4		A498	66.8
	MDA-MB-231	41.9		ACHN	27.5
	MDA-MB-468	62		CAKI-1	39
	HS 578T	53.8		RXF 393	62.9
	MDA-N	22.5		SN12C	29.5
	BT-549	53.9		TK-10	51.3
	T-47D	45.5		UO-31	41.7

minimum quality assurance standards like adaptable to a single suitable growth medium, demonstrate reproducible results for growth inhibition and thereby drug sensitivity under certain laboratory conditions. These cancer cells keep dividing and growing over time, but primary cells of more than 20 passages are not employed in the drug screening process [19]. The cells are exposed to five different doses of drug candidates for 48 h. If they exhibit any of the unique features like (1) preferentially kills any one or more of the tumor cell lines, or (2) acts via a distinctive mechanism, or (3) inhibits tumor growth or malignant cell proliferation at a very small concentration, then the screening will continue to the next stage. In approximately 2500

compounds tested, only 2% may be recommended to proceed to the next
stage of testing in mice.

In vivo *screening using the hollow-fiber technique*: Small hollow plastic, poly-
vinylidene fluoride tubes of 1 mm diameter and 2 cm length containing
human tumor cells are inserted into the body cavity or underneath the skin
of the mouse. The average duration of this test is about 4 days, wherein the
candidate drug is administered to the mice in two dosages and is tested
against 12 tumor cells in different hollow fibers [20–22]. Among 150–200
compounds tested by this method in a year, those molecules that retard cell
growth are carried forward to the next level of screening.

In vivo *testing using xenografts*: Candidate drugs that display positive evidence
of activity in the hollow fiber method are selected for testing in xenografts.
Human tumors are injected directly under the skin or into the peritoneal
cavity of mice, and the drugs are administered at different sublethal dosages. The average duration of this test is about 30 days [23,24]. Compounds
that inhibit or retard tumor growth and exhibit minimal toxic profiles in the
animal will advance to the subsequent phase of screening.

Formulation, pharmacology and toxicology studies: Drug candidates that
reach the above *in vivo* phase of testing are under severe consideration for
human trials so that significant capitals may be committed in due course for
their development [25]. At the initial level of the study, the basic pharmacology of the drug candidate in animals is determined to identify the site of
drug metabolism. The best drug formulation for administration with appropriate dosage, the time interval between each dose administration, and
the mode of drug intake is also established. After these initial metabolic
studies, if the test molecule progresses to the next level, NCI releases the
necessary resources for its further development. In addition, a large-scale
production plan can also be designed for the test compound if required. In
the second level, toxicology studies are performed in two animal species
with the same formulation under consideration for trial administration in
humans. If the drug passes the trial with no severe problems, the dose,
schedule, and route of administration for early phase I clinical trials in cancer patients are suggested.

Clinical trials: Drug clinical studies in humans can commence only after
review and approval by the Food and Drug Administration (FDA) as well
as a local Institutional Review Board (IRB), which consists of a panel of
eminent scientists and nonscientists from research institutions and hospitals
that supervise clinical research [26].

Subsequent to the micro-dosing or Phase 0 trial which bridges the gap between
preclinical and clinical studies, Phase I trials are initiated. In humans, clinical development of anticancer chemotherapeutic agents reasonably follows mainly three
phases. Phase I trials involve the initial administration of either a new drug or drug
combination or hybrid drug to human beings. The primary objective is to decide
the appropriate Phase II drug dose and also to gather more insight into the toxicity,
pharmacokinetic, and pharmacodynamic profiles of the drug. The goal of Phase II

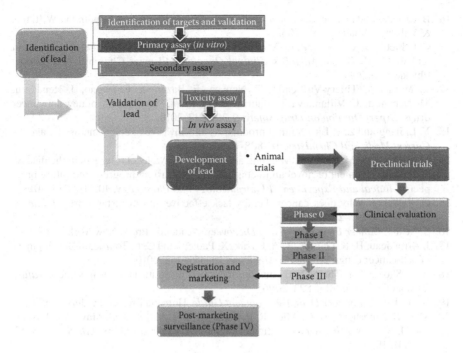

FIGURE 1.1 Schematic representation of various stages in anticancer drug development.

trials is to identify signals of antitumor activity in a specific type of tumor. Phase III trials are designed to relate the efficacy of a new treatment with the current standard of care and can ultimately lead to regulatory approval when positive results are obtained. The various steps involved in the screening of molecules intended to finally qualify as an effective anticancer chemotherapeutic is depicted in Figure 1.1.

REFERENCES

1. V. DeVita, S. Hellman and S. Rosenberg, *Cancer, Principles & Practice of Oncology.* Lippincott Williams & Wilkins, Philadelphia, 2005.
2. D. Hanahan and R. A. Weinberg, The hallmarks of cancer, *Cell,* 100, 57–70, 2000.
3. G. Weber, Why does cancer therapy lack effective anti-metastasis drugs? *Cancer Letters,* 328(2), 207–211, 2013.
4. J. Niederhuber, J. Armitage, J. Doroshow, M. Kastan and J. Tepper, *Abeloff's Clinical Oncology Review.* Elsevier Health Sciences, London, 2013.
5. L. Goodman, A. Gilman, L. Brunton, J. Lazo and K. Parker, *Goodman & Gilman's the Pharmacological Basis of Therapeutics.* McGraw-Hill, New York, 2006.
6. J. Abraham, J. Gulley and C. Allegra, *Bethesda Handbook of Clinical Oncology.* Lippincott Williams & Wilkins, Philadelphia, 2010.
7. M. Perry, *The Chemotherapy Source Book.* Wolters Kluwer Health/Lippincott Williams & Wilkins, Philadelphia, 2008.
8. R. Souhami and J. Tobias, *Cancer and Its Management.* Blackwell Publishing, Malden, 2005.
9. C. Almeida and S. Barry, *Cancer.* Wiley-Blackwell, Chichester, UK, 2010.

10. B. Chabner and D. Longo, *Cancer Chemotherapy and Biotherapy*. Lippincott Williams & Wilkins, Philadelphia, 2006.
11. C. E. Freter, M. C. Perry, M. D. Abeloff, J. O. Armitage, J. E. Niederhuber, M. B. Kastan and W. G. McKenna, *Abeloff's Clinical Oncology*. Elsevier Churchill Livingstone, Philadelphia, 2008.
12. L. Mansi, A. Thiery-Vuillemin, T. Nguyen, F. Bazan, F. Calcagno, J. Rocquain, M. Demarchi, C. Villanueva, T. Maurina and X. Pivot, Safety profile of new anticancer drugs, *Expert Opinion on Drug Safety*, 9, 301–317, 2010.
13. Y. L. Jiang and Z. P. Liu, Natural products as anti-invasive and anti-metastatic agents, *Current Medicinal Chemistry*, 18, 808–829, 2011.
14. M. S. Wong, S. M. Sidik, R. Mahmud and J. Stanslas, Molecular targets in the discovery and development of novel antimetastatic agents: Current progress and future prospects, *Clinical and Experimental Pharmacology and Physiology*, 40, 307–319, 2013.
15. G. F. Weber, Why does cancer therapy lack effective anti-metastasis drugs? *Cancer Letters*, 328, 207–211, 2013.
16. S. Neidle, *Cancer Brug Design and Discovery*. Academic Press, New York, 2008.
17. J. Arrondeau, H. K. Gan, A. R. A. Razak, X. Paoletti and C. L. Tourneau, Development of anti-cancer drugs, *Discovery Medicine*, 10, 355–362, 2010.
18. R. H. Shoemaker, The NCI60 human tumour cell line anticancer drug screen, *Nature Reviews Cancer*, 6, 813–823, 2006.
19. B. Teicher, *Anticancer Drug Development Guide*. Humana Press, New Jersey, 1997.
20. M. G. Hollingshead, M. C. Alley, R. F. Camalier, B. J. Abbott, J. G. Mayo, L. Malspeis and M. R. Grever, *In vivo* cultivation of tumor cells in hollow fibers, *Life Sciences*, 57, 131–141, 1995.
21. Q. Mi, J. M. Pezzuto, N. R. Farnsworth, M. C. Wani, A. D. Kinghorn, and S. M. Swanson, Use of the *in vivo* hollow fiber assay in natural products anticancer drug discovery, *Journal of Natural Products*, 72, 573–580, 2009.
22. O. Temmink, H. Prins, E. Van Gelderop and G. Peters, The hollow fibre assay as a model for *in vivo* pharmacodynamics of fluoropyrimidines in colon cancer cells, *British Journal of Cancer*, 96, 61–66, 2007.
23. J. Jung, Human tumor xenograft models for preclinical assessment of anticancer drug development, *Toxicological Research*, 30, 1–5, 2014.
24. A. Richmond and Y. Su, Mouse xenograft models vs GEM models for human cancer therapeutics, *Disease Models and Mechanisms*, 1, 78–82, 2008.
25. M. Hidalgo, S. G. Eckhardt, N. J. Clendeninn, and E. Garrett-Mayer, *Principles of Anticancer Drug Development*. Springer Science & Business Media, New York, 2010.
26. P. Gupta, V. Gupta and Y. K. Gupta, Phase I clinical trials of anticancer drugs in healthy volunteers: Need for critical consideration, *Indian Journal of Pharmacology*, 44, 540, 2012.

2 *In Vitro* Biological Assays for Anticancer Chemotherapeutics

2.1 INTRODUCTION

In vitro bioassays are any qualitative or quantitative analysis that is performed in a sterile laboratory environment to measure the activity profile of a drug on a live tissue cell or in a living organism [1]. The three essential components of *in vitro* bioassays are (1) stimulus like drug candidates, (2) substrate like tissues, cells, biochemicals, and so on, and (3) response of the substrate to various doses of stimulus. The highly time-consuming and complex process of bringing a new drug to market can be made a lot simpler and effective using *in vitro* assays in the initial stages of drug development. These assays considerably cut down the time taken to eliminate an ineffective drug by observing the interaction between the drug and the cells quickly. *In vitro* assays thereby allow scientists to determine whether the drug has the desired effect and focus their attention solely on those promising molecules that may prove successful and can then progress to the next phases of testing [2].

Use of standards plays a significant role in *in vitro* bioassays. The results of the assay need to be validated by monitoring and comparing the effect with an available known compound (standard). Without judicious choice of standard and its reproducible results in an assay system, no screening can be claimed credible.

2.2 ADVANTAGES OF *IN VITRO* ASSAYS

Though *in vitro* testing still requires subsequent *in vivo* and clinical phases, there are ample supports in using *in vitro* systems to improve understanding on the mechanistic aspects of drug interactions, and the use of human cell lines and tissues to determine toxic effects specific to humans. Promising results obtained during the *in vitro* tests can indicate to some extent whether a new drug will be effective on human patients. The several salient features or advantages in employing *in vitro* screening studies are as follows [3,4]:

- Reduced use of animals
- Predictive potential
- Rapid and speedy process of testing new drugs
- Reduce overall expenditures of drug development

- Better control over the biochemical reactions
- Reproducible results
- Possible to carry out cell based or mechanistic investigation based studies
- High-throughput or able to process and screen a large number of compounds
- Requires only a small quantity of sample for testing
- Possible to perform tests on human blood samples that can be replenished or on tissue samples collected from cadavers
- Permits the use of human tissue samples, which allow a better understanding of drug effect on human cells
- Allows testing the ability of compounds to prevent cell proliferation or induce cell death by taking advantage of various cell properties
- Enables using the outcome of the study to fix an appropriate dose during preclinical *in vivo* studies

2.3   DISADVANTAGES OF *IN VITRO* ASSAYS

Though the quest for alternatives to animal testing by *in vitro* models has gained a new momentum due to regulatory constraints and ethical considerations, there are many associated drawbacks also for *in vitro* assays [5,6].

- Difficulty in maintenance of cell cultures
- Risk of getting negative results with those compounds that get activated after body metabolism or vice versa
- No pharmacokinetic information

2.4   CLASSIFICATION OF *IN VITRO* ASSAYS

The present translational research chain employed in the anticancer drug development process is generally visualized as almost stepwise, categorized system of initial *in vitro* studies and subsequent animal models converging toward final clinical trials and ultimately to patient care standards [7]. Standardized and well-established *in vitro* techniques are designed for the experimental assessment of efficacies of new anticancer agents and these assays allow reduction in the number of promising agents for further clinical testing. The primary *in vitro* bioassays can fall into any of the following categories: nonphysiological assays, biochemical- or mechanism-based assays, cell-based or tissue-based bioassays. The functional assays measure survival of tumor cells both with and without therapy, either as total number of cells or as number of colonies, whereas nonfunctional assays are often employed for mechanistic investigations like assessment of drug effects on apoptotic pathways or intracellular signaling. Both functional and nonfunctional assays are essential for the evaluation of various features of anticancer agents and to improve our understanding of the underlying mechanisms of action of drugs [8–10]. The selection of an assay is mainly decided on the type of study one requires.

2.5 ESSENTIAL TECHNIQUES OF CELL CULTURES

Cell culture techniques employ a number of key elements and features that are universally applied, although there can be diverse ways in practicing these procedures.

2.5.1 PRIMARY CELL CULTURES

A primary cell culture is the initial culture set up directly from a body tissue such as primary or metastatic solid tumor or cell suspensions. Cell suspensions are convenient to develop primary cultures as they grow as single cells or clusters and avoid mechanical or enzymatic dispersion. Cancer cells differ from most normal cell types in their ability to grow in suspension, but generally cultures are initiated by allowing cells to adhere to a substrate before proliferating.

2.5.2 MAINTENANCE OF CELL LINES

Cell cultures should be regularly monitored for the presence of contaminants, variations in pH, and the cell morphology and density should be examined by microscope. Media plus serum and other additives should be changed routinely. The frequency of media renewal will be dependent on the growth rate of the culture.

2.5.3 SUBCULTURE OF CELLS

The adherent monolayer is dislodged (harvesting) from the tissue culture flasks when the culture has occupied the complete surface of the flask, using proteolytic enzymes such as trypsin (trypsinization) to achieve a single cell suspension (subculture or passaging) to maintain healthy cell lines. The culture media is aspirated out from a 60%–70% confluent flask using a micropipette. The cells are washed with phosphate buffered saline (PBS) to eliminate trace amounts of media. Trypsin-EDTA solution is added to each cell culture flask and it is aspirated after a few seconds and the flask is incubated for 3–4 min for cell detachment. Trypsin activity is ceased by adding media containing 10% fetal bovine serum (FBS) after ensuring complete detachment of cells. The determination of cell number while setting up experiments with cancer cell lines can be performed using a hemocytometer or automated Beckman Coulter.

REFERENCES

1. V. Gonzalez-Nicolini and M. Fussenegger, *In vitro* assays for anticancer drug discovery—A novel approach based on engineered mammalian cell lines, *Anti-Cancer Drugs*, 16, 223–228, 2005.
2. M. M. Lieberman, G. M. Patterson and R. E. Moore, *In vitro* bioassays for anticancer drug screening: Effects of cell concentration and other assay parameters on growth inhibitory activity, *Cancer Letters*, 173, 21–29, 2001.
3. D. Zips, H. D. Thames and M. Baumann, New anticancer agents: *In vitro* and *in vivo* evaluation, *In Vivo*, 19, 1–7, 2005.

4. B. A. Teicher, *Anticancer Drug Development Guide: Preclinical Screening, Clinical Trials and Approval.* Springer Science & Business Media, 2013.

5. R. H. Shoemaker, The NCI60 human tumour cell line anticancer drug screen, *Nature Reviews Cancer*, 6, 813–823, 2006.

6. D. R. Ferry and D. J. Kerr, Mechanistic approaches to phase I clinical trials, *Anticancer Drug Development*. Academic Press, USA, pp. 371–381, 2002.

7. M. Baumann, S. M. Bentzen, W. Doerr, M. C. Joiner, M. Saunders, I. F. Tannock and H. D. Thames, The translational research chain: Is it delivering the goods? *International Journal of Radiation Oncology Biology Physics*, 49, 345–351, 2001.

8. J. McCauley, A. Zivanovic and D. Skropeta, Bioassays for anticancer activities, *Metabolomics Tools for Natural Product Discovery: Methods and Protocols*, Springer Science & Business Media, Germany, vol. 1055, pp. 191–205, 2013.

9. S. Neidle, *Cancer Drug Design and Discovery*. Academic Press, USA, 2011.

10. V. Jurisic and V. Bumbasirevic, *In vitro* assays for cell death determination, *Archive of Oncology*, 16(4), 49–54, 2008.

3 Cell Viability and Cytotoxicity Assays

3.1 INTRODUCTION

Cell viability denotes the number of healthy cells in culture, but cannot distinguish between the cells that are actively dividing and those in the quiescent phase. Cytotoxicity of any sample is the quality of being toxic to cells and is expressed in terms of half-maximal inhibitory concentration (IC_{50}) values. *In vitro* cell-based assays have been developed for quick and precise assessment of the cytotoxic activity of test compounds by determining viable cell counts. They also aid in identifying the dissimilarities in the susceptibility of different target cells toward various chemotherapeutic agents. *In vitro* testing approaches in toxicological studies are useful in evaluating the cytotoxic, carcinogenic, and mutagenic effects of various chemical compounds on human cells [1].

Cell viability, membrane integrity, and DNA content are among the most specific and sensitive parameters for measuring cytotoxicity [2]. The cells are treated with different concentrations of the test drug, and the response is examined, and the data are recorded using a plate reader to assess cell viability. These simple assay methods are not designed for any high content imaging. The results received from different assays are centered on quantification of a marker activity related to viable cells and a variety of cellular functions such as enzyme activity, ATP generation, co-enzyme production, cell membrane permeability, cell adherence, nucleotide uptake activity, and so on [3]. This chapter provides an overview of the frequently used cytotoxicity assay techniques to estimate cell viability.

3.2 CELL MEMBRANE INTEGRITY ASSAYS

Cells that are exposed to compounds that possess cytotoxic effects often compromise their cell membrane integrity. Hence, one of the most frequent and direct ways to assess cytotoxic nature of a test sample and to measure cell viability is to measure the cell membrane integrity [4]. Most of the current assays that evaluate the cytotoxicity of a compound are focused on either variation in plasma membrane permeability and subsequent release or leakage of intracellular components or the cellular uptake of dyes generally excluded by healthy cells with uncompromised membrane integrity and further staining of the intracellular components. The main drawback of such permeability assays relies on the fact that the early sites of damage of several cytotoxic agents are intracellular. This irreversible damage of the cell can lead to cell death with an intact plasma membrane [5]. Despite the fact that these tests tend to misjudge cellular impairment when paralleled to other approaches, some

11

permeability assays are still widely accepted and are being used for preliminary cell viability/cytotoxicity measurements.

3.2.1 DYE EXCLUSION ASSAYS

These assays rely on the structural integrity of cells [6].

3.2.1.1 Trypan Blue Assay

Trypan blue (TB) exclusion assay is one of the most frequently employed procedures for evaluating cell viability. The TB molecule (Figure 3.1) is cell membrane impermeable, and hence can only enter into cells having compromised cell membranes [7]. This test is based on a visual examination to determine whether cells in a suspension mixed with TB dye take up or exclude the dye. A viable cell, which is often small, round, and refractive, will have a clear cytoplasm opposed to a nonviable cell that will be swollen, larger, and dark blue with a blue cytoplasm. Inside the nonviable cell, TB binds to intracellular proteins and gives the cells a bluish color under a light microscope. Thus, this assay allows distinguishing live (exclude TB dye and hence unstained) and dead (stained blue as they lose membrane integrity) cells in a given culture sample (Figure 3.2).

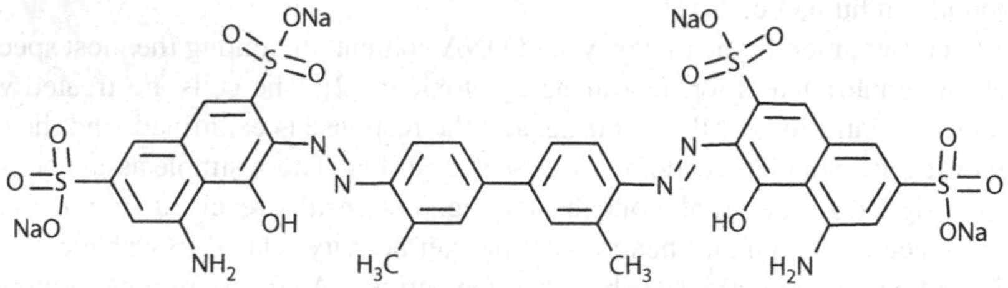

FIGURE 3.1   Structure of Trypan blue dye.

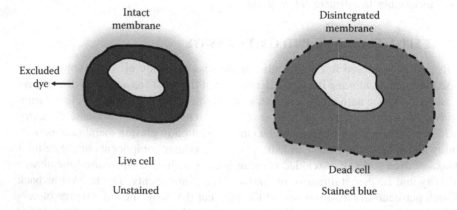

FIGURE 3.2   Trypan blue stained nonviable and unstained viable cells.

Procedure:

1. Trypsinize the monolayer cell culture and adjust the cell population to 1.0×10^5 cells/mL using MEM/DMEM medium containing 10% NBCS (newborn calf serum)/FBS (fetal bovine serum).
2. To each well of the 96-well microtiter plates, add 0.1 mL of the diluted cell suspension (approximately 10,000 cells).
3. After 24 h of incubation, when a partial monolayer is formed, flick off the supernatant, wash the monolayer once with medium, and add 100 µL of different test concentrations prepared using DMEM/MEM with 2% NBCS/FBS (maintenance medium) to the cells.
4. Incubate the plates at 37°C for 3 days in 5% CO_2 atmosphere, and carry out microscopic examination and record the observations every 24 h.
5. After 72 h, discard the test solutions in the wells and trypsinize the cells.
6. Collect the cells by trypsinizing and pellet them by spinning at 1000 rpm for 10 min.
7. Suspend the cell pellet in PBS or maintenance media.
8. Mix the cell suspension with equal volume of 0.4% TB solution.
9. Incubate for 1 min at room temperature.
10. Count the colored cells under microscope or spectrophotometrically at 590 nm or load the cells in a hemocytometer and record the viable cells.

The percentage growth inhibition is calculated using the following formula:

$$\text{Growth inhibition } (\%) = 100 - \frac{(\text{Total cells} - \text{Dead cells})}{\text{Total cells}} \times 100$$

Advantages:

- Quick, simple, and inexpensive method.
- Only small sample size required.

Limitations:

- Cell viability determined indirectly from cell membrane integrity.
- Possibility of a cell's viability being compromised in terms of its capacity to grow or function, maintaining its membrane integrity momentarily.
- The cell may repair itself and become fully viable though it displays abnormal cell membrane integrity transiently.
- Viable as well as nonviable cells may start to take up the dye, on extended periods of TB exposure (>30 min).
- Minor quantities of dye uptake indicating a cell injury may go unobserved, as dye uptake is assessed subjectively.
- Throughput is inadequate for HTS (high-throughput screening) applications.

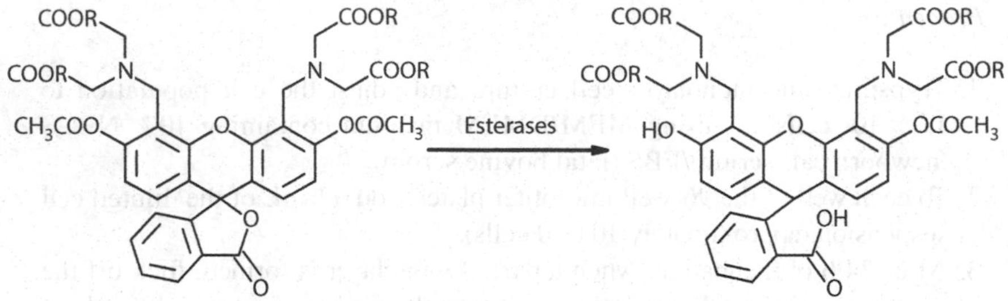

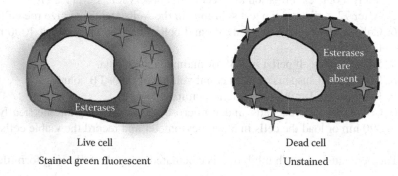

FIGURE 3.3   Reduction of calcein-AM by viable cells to green fluorescent calcein.

Live cell

Stained green fluorescent

Dead cell

Unstained

FIGURE 3.4   Viable cells emit green fluorescence, whereas nonviable cells deprived of active esterases do not produce a fluorescent green signal.

3.2.1.2   Calcein Acetoxymethyl (Calcein-AM) Assay

Calcein-AM is a nonfluorescent hydrophobic dye that can be used to assess cell viability. Acetoxymethyl ester of calcein is a lipid-soluble diester frequently used to stain viable cells as it can passively permeate intact cell membranes of live cells. Calcein-AM is hydrolyzed by intracellular esterases into a polar lipid-insoluble anionic intense green fluorescent product calcein (Figure 3.3), which is retained in the cytoplasm of cells with intact membranes, but is released by ones with damaged cell membranes. This assay is based on the principle that nonviable cells deprived of active esterases do not produce a fluorescent green signal (Figure 3.4). Hence, release of calcein in the supernatants recovered from cells grown in black-walled plates can be easily quantified in less than 2 h by measuring the fluorescence emitted by live cells with excitation and emission at 490 and 520 nm, respectively. Alternatively, the lytic activity can also be assessed by evaluating the fluorescence retained in healthy cells after quenching the fluorescence released by dead cells [8].

Procedure:

1. Seed around 2×10^6 cells/mL in well plates and allow test drug treatment for desired period.

2. Remove media, trypsinize and add 15 µM calcein-AM (1-mg/mL solution in dry dimethyl sulfoxide) and incubate for 30 min at 37°C with intermittent shaking.
3. Wash with medium, adjust the cells to 10^4/mL.
4. Incubate in 5% CO_2 for 4 h at 37°C.
5. Harvest 75 µL of each supernatant and transfer into fresh plates.
6. Samples are measured using a spectrofluorimeter.

Advantage:

- Simple, rapid, and accurate assay.

Limitations:

- Appropriate cell culture media should be selected as the presence of Fe^{3+}, Cu^{2+}, Ni^{2+}, Co^{2+}, and Mn^{2+} quench the green fluorescence emitted by calcein at physiological pH.
- Esterase activity of serum used to supplement the culture medium can affect the readings.
- Some multidrug resistant cell types can actively pump out calcein-AM.
- Not used in HTS assays owing to liquid handling steps needed to remove nonspecific esterase activity for optimal performance.

3.2.1.3 Propidium Iodide (PI)/7-Aminoactinomycin D (7-AAD) Staining Assay

PI and 7-AAD are membrane impermeable fluorescent dyes, generally excluded from viable cells. They penetrate the membrane of dying or dead cells on membrane damage and bind to nucleic acid by intercalating between the base pairs. On entering the dead cells, PI (Figure 3.5) will intercalate stoichiometrically into double-stranded DNA or double-stranded RNA, while 7-AAD (Figure 3.5) will insert only between base pairs in guanine-cytosine rich regions of double-stranded DNA. The dead cells will remain labeled for data acquisition as the intercalation is mediated by noncovalent forces and these dyes remain in the buffer used to suspend cells. The red fluorescence radiated by PI in dead cells (Figure 3.6) can be quantified either by FACS (fluorescence-assisted cell sorting or flow cytometry) analysis or by

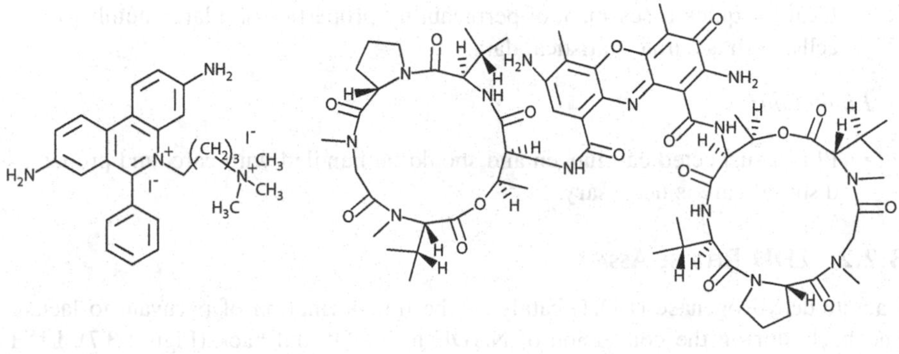

FIGURE 3.5   Structure of propidium iodide (left) and 7-AAD (right).

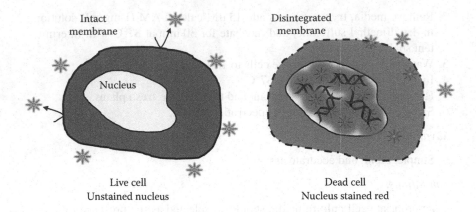

Intact membrane   Disintegrated membrane

Nucleus

Live cell
Unstained nucleus

Dead cell
Nucleus stained red

FIGURE 3.6 Viable cell membrane is impermeable to PI dye and remains unstained. In nonviable cells, PI intercalates into the DNA to give a red fluorescence.

fluorescence microscopy. PI has an excitation maximum at 535 nm and fluorescent emission maximum at 617 nm. 7-AAD can be excited at 488 nm with an argon laser.

Procedure:

1. Seed around 2×10^6 cells/mL in well plates, allow test drug treatment, and incubate for the desired duration.
2. Wash the cells twice with 2 mL PBS, trypsinize the cells, centrifuge the suspension, and decant the buffer from pelleted cells.
3. Resuspend cells in a suitable volume of flow cytometry staining buffer and add 5 µL of PI (1 mg/mL or 1.5 mM) or 7-AAD staining solution for every 100 µL of cells.
4. Incubate for 5–15 min at room temperature or on ice in the dark and analyze cells in a flow cytometer.

Advantages:

• Quick, simple, reliable, and inexpensive method.
• A small population of cells is sufficient to carry out the experiment.
• Ideal for quick assessment of permeability properties of a large number of cells, with accurate statistical data.

Limitation:

• PI is a suspected carcinogen and should be handled with care, and proper disposal care is necessary.

3.2.2 LDH Release Assays

Lactate dehydrogenase (LDH) catalyzes the transformation of pyruvate to lactate and back, during the conversion of NADH to NAD$^+$ and back (Figure 3.7). LDH release to the surrounding media is a marker of cell integrity, and its discharge is

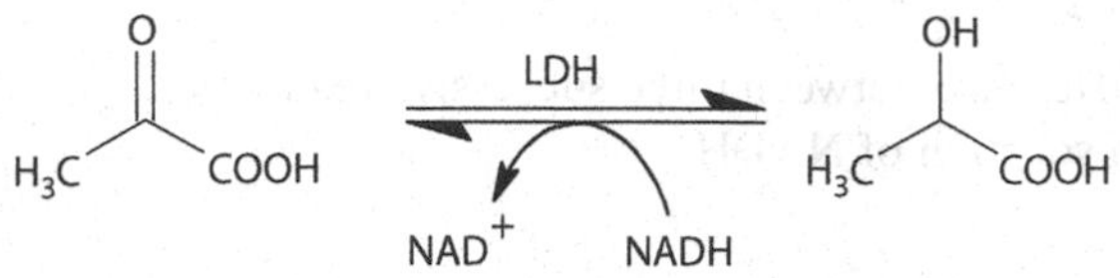

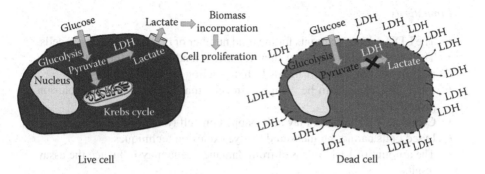

FIGURE 3.7 Pyruvate + NADH + H⁺ ⇌ NAD⁺ + lactate.

FIGURE 3.8 In viable cells, normal LDH activity is observed whereas in nonviable cells, loss in cell membrane integrity releases LDH into an extracellular medium which can be quantified.

possible only when the membrane damage has occurred (Figure 3.8). The LDH release assay is a widely used indirect method for enzyme quantification which is based on the evaluation of LDH activity in the extracellular medium. LDH activity is spectrophotometrically measured in the culture medium and the cellular lysates at 340 nm as cytotoxicity endpoint which correlates with cell membrane integrity and in turn cell viability [9,10].

Procedure:

1. Seed around 2×10^6 cells/mL in well plates and allow test drug treatment for desired period.
2. Prepare PBS of pH 7.4 mM (13.97 g K_2HPO_4 and 2.69 g KH_2PO_4/1000 mL), NADH (2.5 mg of NADH dissolved in 1 mL of phosphate buffer), and 2.5 mg/mL of sodium pyruvate in buffer.
3. Remove the test drug-treated cultures from incubator and transfer ~100 μL culture medium into a micro-centrifuge tube and spin down at 13,000 rpm for 1 min to deposit the cell debris.
4. Transfer 0.1 mL cultured medium into test tube and add 2.4 mL PBS and 0.1 mL of NADH.
5. Incubate for 30 min at room temperature, add 0.1 mL sodium pyruvate solution, and mix well.
6. Measure the absorbance at 340 nm spectrophotometrically. Absorbance is monitored at 60 s interval for 6 min.

$$\text{Serum LDH activity} = \frac{\Delta E\ 340/\text{min} \times 1000 \times \text{Volume in cuvette (mL)}}{\text{Volume of serum} \times 6.3}$$

where:

ΔE = mean difference between three successive readings

6.3 = molar absorption of NADH

Advantage:

• Widely used and accepted method.

Limitations:

• The LDH release accounts for the total number of both dead and lysed cells.
• The LDH stability can vary significantly, extending from the loss of a few percent per day to a half-life of 12 h depending on various cell types.
• The LDH release can be complete in cells marked dead by dye exclusion procedures.
• Complete LDH release may only happen on cell lysis.
• Lysed cells cannot be measured by dye exclusion techniques.
• The amount of LDH released from damaged cells may influence the assay results.

3.3 FUNCTIONAL OR METABOLIC ASSAYS

Metabolic activity is as an indication of cell viability and the metabolic assays measure the vital functions characteristic of viable cells. Cellular damage will inevitably affect the metabolic function and growth of the cell [11]. Most of the universally used cell viability analyses are based on the principle that during cell death, there will be a rapid loss in their capability to transform the substrate to its product. Incubation of the substrate with a cell population will produce a signal that is proportionate to the amount of healthy cells present. The tetrazolium and resazurin reduction assays involve incubation of the reagent with a population of viable cells to convert the substrate to a fluorescent or colored product which can be detected using a plate reader and is based on general cellular metabolism or an enzymatic action as an indicator of healthy cells. Whereas in the ATP assay, instant cell rupture occurs on addition of assay reagent, therefore there is no need for incubation of the reagent with a healthy cell population.

3.3.1 TETRAZOLIUM REDUCTION ASSAYS

The colorimetric assessment of colored formazan product formed by the cleavage of tetrazolium salt as a result of cellular mitochondrial metabolic activity in viable cells (Figure 3.9) forms the basis of these assays [12]. Dead cells are inevitably unable to offer energy for metabolic cell function and hence do not metabolize various tetrazolium compounds. Different types of tetrazolium salts that have been used to quantify viable cells fall into two basic classes: (1) positively charged MTT that penetrates readily into the live eukaryotic cells, and (2) negatively charged XTT, MTS, and WST-1 that do not easily enter the cells and are generally employed together with an intermediate electron acceptor that transfers electrons from the plasma membrane or

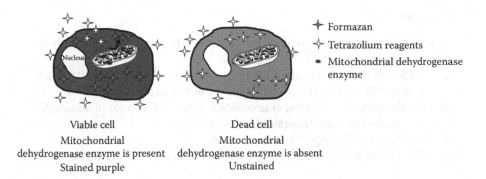

Viable cell
Mitochondrial
dehydrogenase enzyme is present
Stained purple

Dead cell
Mitochondrial
dehydrogenase enzyme is absent
Unstained

† Formazan
† Tetrazolium reagents
* Mitochondrial dehydrogenase
enzyme

FIGURE 3.9 Mitochondrial dehydrogenase enzyme converts tetrazolium dyes to purple formazan in viable cells and are stained purple. Nonviable cells are unstained.

cytoplasm to enable the conversion of tetrazolium dye into its formazan product [13]. The main drawback of these colorimetric assays is that they may misinterpret cellular damage and perceive cell death only toward the final stages of cell death when the metabolic function is reduced. However, the colorimetric assays are of high value in quantifying factor-induced cytotoxicity within a period of 24–96 h of cell culture.

3.3.1.1 MTT Assay

The MTT assay is one of the most extensively used and highly reliable colorimetric assays for estimating the viability of cells that are treated with synthetic or natural products which give an indication of their efficacy to induce cell cytotoxicity [14,15]. MTT cell viability assay is the first best-known enzyme-based investigation developed for determining mitochondrial dehydrogenase activities in the live cells. The assay relies on the ability of mitochondrial succinate dehydrogenase enzymes present in healthy cells to cleave the tetrazolium ring and transform the yellow water-soluble 3-(4,5-dimethylthiazol-2-yl)-2,5-diphenyl tetrazolium bromide (MTT) substrate into purple water-insoluble formazan product within the cells [16,17] (Figure 3.10). These needle-shaped formazan crystals can be dissolved using isopropanol or dimethyl sulfoxide or an acidified ethanol solution. The number of viable cells is proportional to the color intensity of the dissolved formazan and can be quantified spectrophotometrically by using an ELISA plate reader.

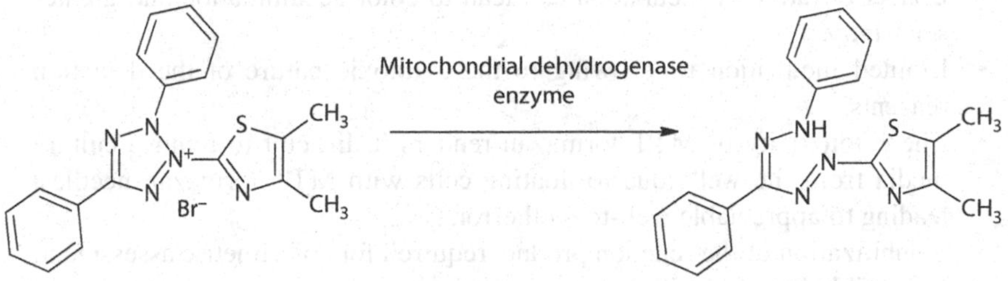

Mitochondrial dehydrogenase
enzyme

FIGURE 3.10 Reduction of MTT to purple formazan.

Procedure:

1. Harvest exponentially growing 1×10^5 cells/mL from T-25 tissue culture flask and prepare a stock cell suspension.
2. Seed 5×10^3 cells in 0.1 mL of MEM medium supplemented with 10% FBS in a sterile 96-well flat bottom tissue culture plate and allow to grow for 24 h.
3. Treat the cells with different concentration of test compounds in quadruplicates and incubate for the desired duration of time.
4. Treat the control group cells with medium.
5. Add 30 µL of MTT reagent (4 mg/mL) into the wells and incubate for 3–4 h at 37°C.
6. Remove medium containing MTT after incubation, and dissolve the formazan crystals in each well in 100 µL of DMSO or n-propanol.
7. Measure the absorbance at 560 nm using an ELISA plate reader.

$$\text{Growth inhibition } (\%) = \frac{(\text{Control} - \text{Test})}{\text{Control}} \times 100$$

Advantages:

• Rapid, easy, safe, versatile, sensitive, inexpensive, quantitative, and highly reproducible method.
• Amenable to large-scale screening.
• MTT reduction correlates with cellular protein content and cell number.
• No need for cell transfer as the entire study is completed in a single microplate.
• Can be used with most cell types as MTT is metabolized by many cells.

Limitations:

• MTT concentration in the medium decides the MTT product formation.
• The kinetics and degree of saturation are cell-type dependent.
• Individual cell numbers cannot be quantitated and the results are expressed as a percentage of control absorbance.
• Less effective if cells are cultured in the same media that has supported growth for a few days, which leads to underestimation of results.
• Longer duration of incubation can lead to color accumulation and greater sensitivity.
• Limited incubation time owing to the cytotoxic nature of the detection reagents.
• The cytotoxicity of MTT formazan renders it difficult to remove culture media from the wells due to floating cells with MTT formazan needles, leading to appreciable well-to-well error.
• Solubilization of the reaction product required for colorimetric assessment.
• Impossible to take multiple time points in a single assay.
• Cells with low metabolic activity must be used in high numbers.

3.3.1.2 MTS Assay

Recently developed improved tetrazolium reagents that include MTS, XTT, and the WST series are more advantageous than MTT, as viable cells can reduce these new dyes to generate formazan products that are directly soluble in the culture medium. This aids in the elimination of a liquid handling step while performing the assay making the procedure more convenient [18]. The cell permeability of the tetrazolium salts is restricted due to the negative charge of the formazan product that supports its dissolution in the culture medium. So they are used in conjunction with intermediate electron acceptor agents like phenazine ethyl sulfate (PES) or phenazine methyl sulfate (PMS) which can enter the healthy cells, get reduced to the soluble formazan product in the cytoplasm, and exit the cells (Figure 3.11). The potential toxic property of the intermediate electron acceptors demands optimization of the technique for different cell types as well as individual assay environments [19]. Intermediate electron acceptor renders optimal performance only within a narrow range of concentrations.

Preparation of reagent:

Prepare 2 mg/mL MTS solution by dissolving MTS powder in Dulbecco's PBS to form a clear golden-yellow solution. Add PES powder in MTS solution (0.21 mg/mL) adjusting pH to 6.0–6.5 using 1N HCl.

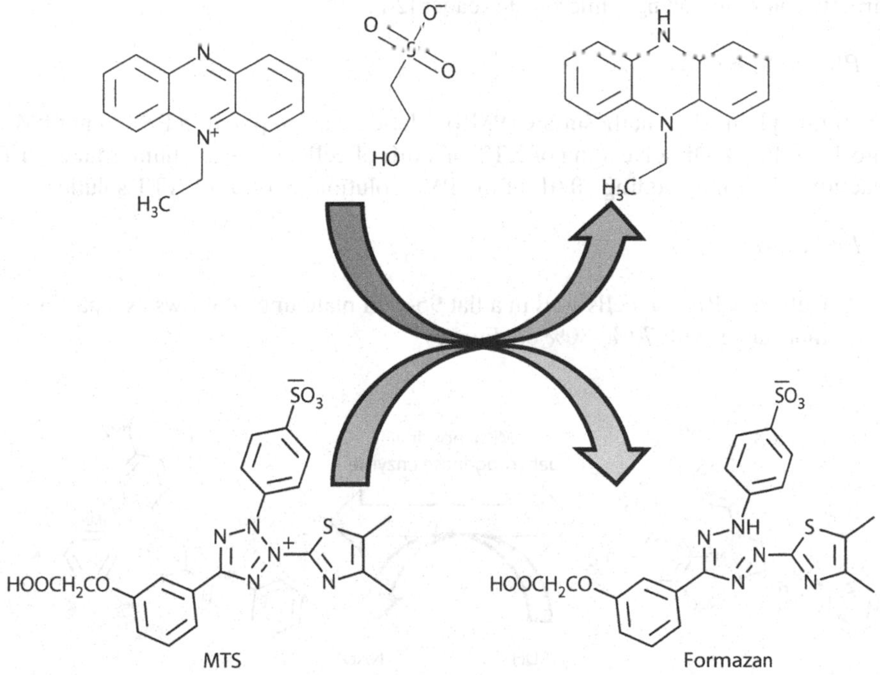

MTS

Formazan

FIGURE 3.11 PES transfers electrons from NADH in the cytoplasm to reduce MTS into formazan.

Procedure:

1. Seed the cells in a 96-well plate to hold a final volume of 100 μL/well.
2. Incubate with test compounds for the desired duration of exposure.
3. Add 20 μL of MTS solution containing PES into each well maintaining the final concentration of MTS as 0.33 mg/mL.
4. Incubate for 1–4 h at 37°C and record absorbance at 490 nm.

Advantages:

- Produce water soluble-formazan product making the assay protocol easy.
- Absorbance can be recorded periodically during initial stages of incubation from the assay plates.

Limitation:

- Results are affected by prolonged incubations with the dye mixture beyond 4 h.

3.3.1.3   XTT Assay

This assay is centered on the cleavage of yellow XTT (2,3-bis-(2-methoxy-4-nitro-5-sulfophenyl]-2*H*-tetrazolium-5-carboxyanilide salt) to form an orange water-soluble formazan dye (Figure 3.12) in metabolically active live cells which can be directly computed using a microplate reader [20].

Preparation of reagent:

10 mM phenazine methosulfate (PMS) solution was prepared in PBS (3 mg PMS into 1 mL PBS). Dissolve 4 mg of XTT in 4 mL of cell culture medium. Make XTT reaction solution by adding 10 μL of the PMS solution to 4 mL of XTT solution.

Procedure:

1. Cultivate 10^4–10^5 cells/well in a flat 96-well plate until it grows as a partial monolayer with 70%–80% confluency.

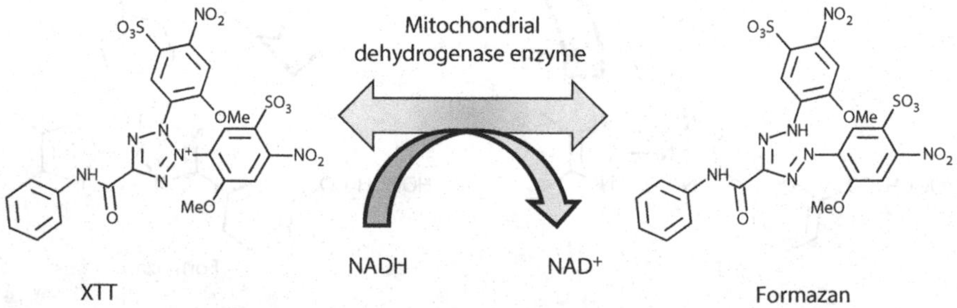

FIGURE 3.12   Reduction of XTT to XTT formazan.

2. Add 100 µL of different concentrations of test drug dilutions prepared in medium per wells in quadruplicates and incubate in a CO_2 incubator at 37°C for the required number of days.
3. Add 25 µL of XTT/PMS solution to each well and incubate further for 2 h.
4. Shake the plate gently to distribute the dye evenly in the wells.
5. Measure the absorbance at 450 nm against a blank background control using an ELISA plate reader.

Advantages:

- Transfer of cells not required, as the entire assay is carried out in a single microplate.
- The reaction product is soluble in water.
- Possible to take multiple time points in a single assay.

Limitations:

- XTT being generally cytotoxic destroys the cells under study and hence allows only a single assessment.
- Presence of PMS required for efficient reduction.
- XTT reagent should be prepared fresh before use.
- All types of cells cannot metabolize XTT.

3.3.1.4 WST Assay

Mitochondrial dehydrogenase enzyme present in the viable cells cleaves WST to aqueous soluble formazan, and forms the basis of water soluble WST assay. WST dyes take up two electrons present in the live cells to produce yellow, orange, or purple formazan dye [21] (Figure 3.13). They are more preferred than MTT dyes mainly due to three reasons: (1) WST-8 (2-(2-methoxy-4-nitrophenyl)-3-(4-nitrophenyl)-5-(2,4-disulfophenyl)-2*H*-tetrazolium) and WST-1 are reduced outside the cells and later combine with PMS electron mediator, (2) they yield a water-soluble formazan that can be read directly without any solubilization step, and (3) the usage of WST gives a better signal and decreases cell toxicity as it does not accumulate within the cells like MTT.

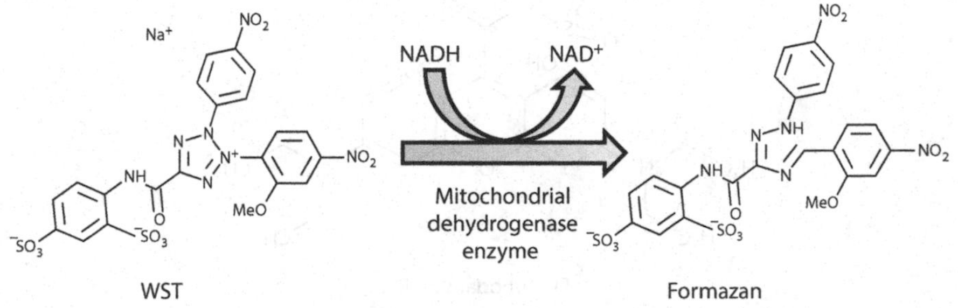

FIGURE 3.13 Reduction of WST to WST formazan.

Procedure:

1. Seed 10^4–10^5 cells/well in a 96-well plate.
2. Incubate with test compounds for the desired period of time at 37°C.
3. Add 10 µL of the reconstituted WST mixture to each well.
4. Mix gently on an orbital shaker for about 1 min to ensure homogeneous distribution of color and incubate the cells for 2 h (adherent culture)–4 h (suspension culture) at 37°C in a CO_2 incubator.
5. Measure the absorbance of each sample using a microplate reader at 450 nm.

Advantages:

• Easier and more simple to perform than MTT assay.
• No need for cell transfer as the whole assay is done in a single microplate.
• The reaction product is water soluble.
• Repeated measurement of the assay is possible.
• Ready-to-use dye solution can be stored in the refrigerator without significant degradation for several weeks.

Limitations:

• Assay is nonlinear over a wide logarithmic cell division range due to the ELISA plate reader.
• All cell types do not metabolize WST-1.

3.3.2   SULFORHODAMINE B ASSAY

Sulforhodamine B (SRB) is a bright pink anionic amino-xanthene dye (Figure 3.14) that binds electrostatically to the basic amino acids of intracellular proteins in mild acidic conditions. It dissociates when exposed to basic conditions and can be

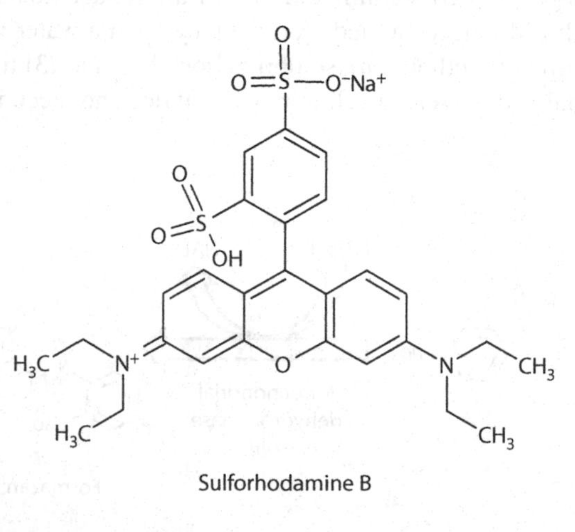

Sulforhodamine B

FIGURE 3.14   Sulforhodamine B dye.

extracted from cells and solubilized for colorimetric measurement to estimate the total protein content. The cellular proteins are quantitatively stained which is proportional to cell number and correlates with overall protein synthesis rate and therefore with cell viability and proliferation [22]. This assay is effectively used to screen cytotoxicity using a large panel of cancer cells derived from various kinds of tumors.

Procedure:

1. Plate 5000–10,000 cells/well in 96-well flat bottom plates and keep them overnight.
2. Once the partial monolayer of cells is formed, treat them with test drug for the desired duration and incubate the plates at 37°C in a CO_2 incubator.
3. Add cold 50% trichloroacetic acid (TCA) to a final concentration of 10% to fix the cells.
4. Incubate for 1 h at 4°C and wash cells five times with deionized water.
5. Stain the cells with 0.4% SRB and dissolve in 1% acetic acid for 15–30 min.
6. Wash five times with 1% acetic acid to remove unbound stain.
7. Air-dry the plates at room temperature and solubilize the bound dye with 10 mm Tris base.
8. Analyze the plates at 595 nm on a microplate reader.

$$\text{Growth inhibition } (\%) = \frac{(\text{Control} - \text{Test})}{\text{Control}} \times 100$$

Advantages:

- Superior method for assessing chemosensitivity due to better linearity and higher sensitivity with cell number.
- A stable end point assay that does not require any time-sensitive measurement.
- Accurate, reproducible, and inexpensive.
- SRB staining is stable and plates can be stored for extended periods ranging from several weeks up to several months as opposed to MTT assay.
- The microscopic pictures of stained test drug-treated wells can give visual proofs of the study.
- The procedure can be interrupted at several steps.

Limitations:

- Needs addition of TCA for cell fixation.
- TCA addition should be gentle or it could dislodge the cells before they become fixed, and can generate incorrect results.

3.3.3 ACID PHOSPHATASE ASSAY

Acid phosphatases (APs) belong to the hydrolase class of enzymes that catalyze the hydrolysis of orthophosphate monoesters in acidic environments (pH = 4.8) [23]. This

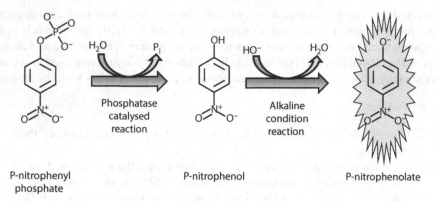

P-nitrophenyl phosphate

P-nitrophenol

P-nitrophenolate

FIGURE 3.15 In viable cells, APs catalyze hydrolysis of *p*-nitrophenol phosphate. Alkali addition gives yellow color, which can be quantified.

enzyme effectively cleaves pyrophosphate (a cellular waste product) and converts it to a usable phosphate. The cell-membrane associated acid phosphatase in live cells transforms colorless P-nitrophenyl phosphate substrate to colorless *p*-nitrophenol product in acidic conditions. The conversion reaction is ceased at the end of 30 min incubation by the addition of 0.05 M NaOH which raises the pH of the reaction medium (Figure 3.15). Para-nitrophenol is yellow at alkaline pH, and its concentration can be measured spectrophotometrically [24]. The rate of increased absorbance at 405 nm is proportional to the AP enzyme activity and concomitant change in the amount of substrate converted, which in turn is proportional to the amount of viable cells indicating the degree of cytotoxicity caused by the test material.

Procedure:

1. Cultivate 10^4–10^5 cells/well in a flat 96-well plate until it grows as a partial monolayer with 70%–80% confluency.
2. Incubate with test compound for the desired duration.
3. Remove the culture medium from the wells and wash the cells by adding 200 µL PBS (pH 7.2) and discard the buffer by flipping the plate.
4. Remove medium/buffer from plates by centrifuging the 96-well plates at 2500 rpm for 10 min.
5. Add 100 µL of buffer containing 0.1% Triton® X-100, 0.1 M sodium acetate (pH 5.0), and 5 mM *p*-nitrophenyl phosphate to each well and incubate the plate at 37°C for 2 h.
6. Add 10 µL of 1 N NaOH to arrest the reaction and measure the color intensity at 405 nm using a microplate reader.

Advantage:

- Simple, accurate, and yields reproducible results.

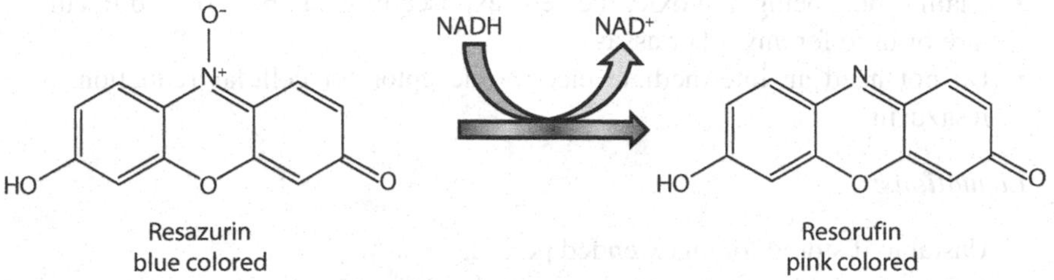

FIGURE 3.16 Reduction of resazurin substrate to resorufin product in viable cells.

3.3.4 ALAMAR BLUE OXIDATION-REDUCTION ASSAY

Alamar blue (AB) is a cell viability assay reagent which contains the cell permeable, nontoxic, and weakly fluorescent blue indicator dye called resazurin (7-hydroxy-3H-phenoxazin-3-one-10-oxide) used to monitor the reducing environment of viable cells. Resazurin undergoes enzymatic reduction due to the action of flavin mononucleotide dehydrogenase, flavin adenine dinucleotide dehydrogenase, nicotinamide adenine dehydrogenase, and cytochromes inside the mitochondria to generate pink-colored and highly fluorescent red resorufin (Figure 3.16). During normal cellular metabolism, the red resorufin is excreted to the medium outside the viable cells which results in visible color change from a completely oxidized nonfluorescent blue to a completely reduced fluorescent red [25]. The rate of reduction based on color change can be quantified colorimetrically or fluorometrically which in turn reflects the number of viable cells.

Procedure:

1. Culture 10^4–10^5 cells/well in a flat opaque-walled (black) 96-well plate until it grows as a partial monolayer with 70%–80% confluency.
2. Incubate with test drug for the desired period of exposure.
3. Prepare 0.15 mg/mL solution of resazurin in Dulbecco's phosphate-buffered saline (DPBS) maintained at pH 7.4.
4. Filter the resazurin solution through a 0.2 μ filter into a light-protected sterile container.
5. Add 20 μL resazurin solutions to each well and incubate at 37°C for 1–4 h.
6. Record the fluorescence using either 560 nm excitation or 590 nm emission.

Advantages:

- Simple, rapid, and relatively cheap assay.
- Requires no lysis, extraction, or washing of sample.
- Faster and less artifact prone than MTT assay.
- Uses a homogeneous format.
- More sensitive than tetrazolium assays.
- Can be combined with other approaches such as caspase activity measurement to collect more evidence about the cytotoxic mechanism.

- Alamar blue being nontoxic, the cells exposed to it can be returned to culture or used for any other assays.
- Do not need an intermediate electron acceptor for cellular reduction of resazurin.

Limitations:

- Unstable if stored for an extended period.
- It shows variable metabolic behavior under different cell culture conditions.
- No standard available and hence needs to be standardized for every cell line under investigation.
- Requires incubation of the substrate with viable cells for an adequate period at 37°C to generate a signal.
- Alamar blue is very sensitive, and the results depend on cell type and incubation time.

3.3.5 NEUTRAL RED ASSAY

Lysosomal integrity, with the concomitant binding of the neutral red dye, is an extremely sensitive indicator of cell viability. The weakly cationic neutral red (3-amino-7-dimethyl-2-methylpnenazine hydrochloride) dye can penetrate the viable cell membranes by nonionic passive diffusion and binds via electrostatic hydrophobic bonds to anionic and/or phosphate groups of the lysosomal matrix (Figure 3.17). The neutral red uptake depends on the live cell's capacity to maintain pH gradients, through ATP production. At physiological pH, the dye holds a net charge close to zero, which enables it to penetrate the cell membrane. There is a proton gradient to maintain a lower pH inside the lysosomes than that of the cytoplasm. This converts the dye to a charged form which can be retained easily inside the lysosomes [26]. The dye is then extracted from the viable cells using an acidified ethanol solution, and the absorbance of the solubilized dye is quantified using a spectrophotometer [27]. Thus, it is possible to differentiate between viable, damaged, or dead cells based on their specific lysosomal capacity to take up the dye.

Procedure:

1. Culture 10^4–10^5 cells/well in a flat opaque-walled (black) 96-well plate until it grows as a partial monolayer with 70%–80% confluency.
2. Incubate the cells for desired period with the test compounds.

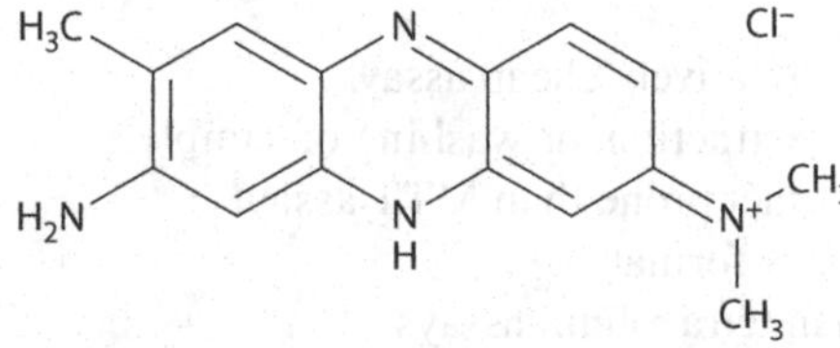

FIGURE 3.17   Neutral red dye.

3. Add 0.33% neutral red solution equal to 10% of the culture medium volume.
4. Incubate for 2–4 h depending on the cell type and maximum cell density to allow neutral red uptake by the cells.
5. Decant excess dye and add solvent to all wells to extract the neutral red dye contained within the cells.
6. Allow the cultures to stand for 10 min at room temperature.
7. Gently stir in a gyratory shaker or gently rock the plate to evenly mix the solubilized dye.
8. Spectrophotometrically measure absorbance at 540 nm.

Advantage:

• Simple, sensitive, quick, and economical method.

Limitation:

• Incubation period should be optimized as it varies with cell densities or cellular metabolic activity.

3.3.6 PROTEASE VIABILITY MARKER ASSAY

Measurement of protease activity conserved within healthy cells serves as a cell viability indicator. The fluorogenic cell permeable protease substrate GF-AFC (glycylphenylalanyl-aminofluorocoumarin) can precisely identify protease activity that is confined to only intact live cells [28]. The GF-AFC penetrates viable cells where cytoplasmic aminopeptidase cleaves the glycine and phenyl alanine residues to liberate aminofluorocoumarin (AFC), which produces a fluorescent signal proportional to the total number of live cells (Figure 3.18). The protease activity wanes rapidly with cell death, making it an ideal selective marker of a viable cell population.

Procedure:

1. Culture 10^4–10^5 cells/well in a flat opaque-walled (black) 96-well plate until it grows as a partial monolayer with 70%–80% confluency.
2. Add test compounds such that the final volume is 100 µL/well.
3. Incubate the cells for the preferred exposure period.
4. Add GF-AFC substrate (10 µL) to assay buffer (10 mL) and mix well to prepare the CellTiter-Fluor™ reagent.

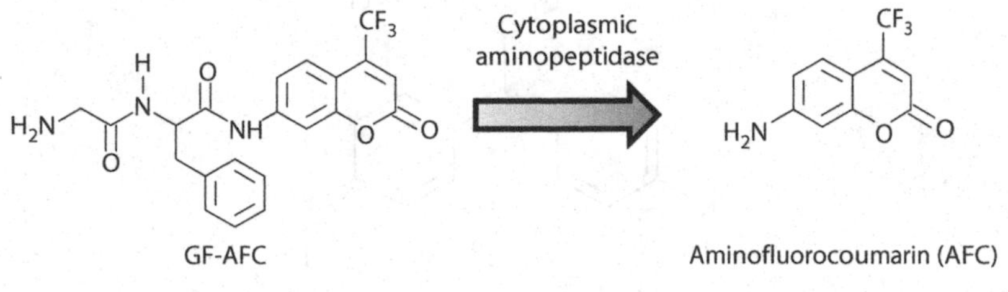

FIGURE 3.18   Conversion of GF-AFC to fluorescent AFC in viable cells.

5. Add 100 µL CellTiter-Fluor reagent to each well and mix by orbital shaking.
6. Incubate the cells for 30 min at 37°C.
7. Measure the fluorescence at 380–400 nm using a fluorometer.

Advantages:

- The signal generated from the protease assay correlates well with other established cell viability methods.
- GF-AFC is less toxic to the cells in culture.
- Extended exposure of the GF-AFC substrate to the cells does not affect cell viability as opposed to tetrazolium or neutral red assays.
- Cell can be further used for subsequent assays after recording fluorescence data from the protease assay.
- Incubation time needed to obtain a satisfactory signal is considerably shorter (30 min–1 h) compared to 1–4 h required for the tetrazolium assays.

Limitation:

- The CellTiter-Fluor viability reagent should be used within 24 h if stored at room temperature.

3.4 CELL ADHESION ASSAY

Adherent cells detach from cell culture plates during cell death. This characteristic can be utilized for the indirect quantification of cell death and to determine differences in proliferation on stimulation with death-inducing agents. One simple method to detect maintained adherence of cells is the staining of attached cells with crystal violet dye (methyl violet 10B), hexamethyl pararosaniline chloride, or pyocyanin as in Figure 3.19, which binds to proteins and DNA. Cells that undergo cell death lose their adherence and are subsequently lost from the population of cells, reducing the amount of crystal violet staining in a culture. The dye uptake by viable cells is measured colorimetrically soon after acetic acid dye elution [28,29].

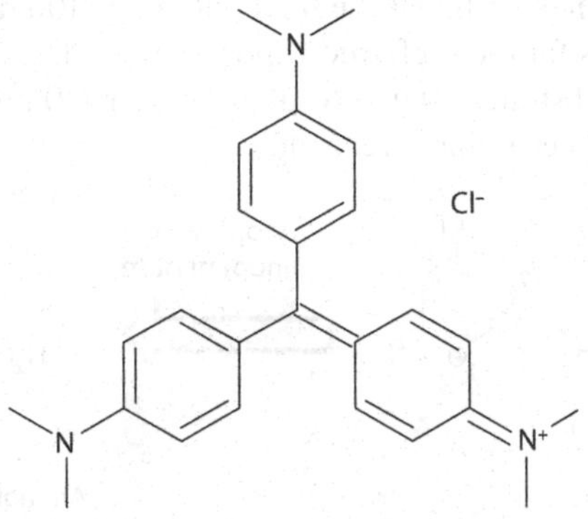

FIGURE 3.19 Structure of crystal violet dye.

Procedure:

1. Seed 5×10^5 cells in 0.1 mL of MEM medium supplemented with 10% FBS in a sterile 96-well flat bottom tissue culture plate and allow cell attachment.
2. Treat the cells with test compounds and incubate for desired period of time.
3. Remove the culture medium from wells and wash the cells twice with PBS.
4. Replace with PBS containing 1% paraformaldehyde (0.25 g dissolved in 25 mL hot PBS) for 15 min at room temperature and rinse twice with PBS.
5. Stain with 0.1% crystal violet (dissolve 0.5 g crystal violet in 80 mL deionized water and add 20 mL of methanol) for 15 min at room temperature.
6. Pour off extra crystal violet solution.
7. Gently wash the cells with deionized water until the water no longer runs dark.
8. Photograph the dishes or scan for colony counts or add 100 μL of 33% glacial acetic acid in each well, and then mix the contents and measure the absorbance at 570 nm.

REFERENCES

1. J. McCauley, A. Zivanovic and D. Skropeta, Bioassays for anticancer activities, In U. Roessner and D. A. Dias (eds.) *Metabolomics Tools for Natural Product Discovery: Methods and Protocols*, Springer Science & Business Media, Germany, vol. 1055, pp. 191–205, 2013.
2. H. Vogel, *Drug Discovery and Evaluation: Pharmacological Assays*. Springer, New York, 2007.
3. D. Zips, H. D. Thames and M. Baumann, New anticancer agents: *In vitro* and *in vivo* evaluation, *In Vivo*, 19, 1–7, 2005.
4. D. Pegg, Viability assays for preserved cells, tissues and organs, *Cryobiology*, 26, 212–231, 1989.
5. M. C. Wusteman, D. E. Pegg, M. P. Robinson, L. H. Wang and P. Fitch, Vitrification media: Toxicity, permeability and dielectric properties, *Cryobiology*, 44, 24–37, 2002.
6. S. A. Altman, L. Randers and G. Rao, Comparison of trypan blue dye exclusion and fluorometric assays for mammalian cell viability determinations, *Biotechnology Progress*, 9, 671–674, 1993.
7. W. Strober, Trypan blue exclusion test of cell viability, *Current Protocols in Immunology*, 111, A3.B.1–A3.B.3, 2015.
8. R. P. Haugland, *The Handbook: A Guide to Fluorescent Probes and Labeling Technologies*. Molecular probes, Invitrogen Corp, Eugene, 2005.
9. S. M. Smith, M. B. Wunder, D. A. Norris and Y. G. Shellman, A simple protocol for using a LDH-based cytotoxicity assay to assess the effects of death and growth inhibition at the same time, *PloS One*, 6, e26908, 2011.
10. B. S. Cummings and R. G. Schnellmann, Measurement of cell death in mammalian cells, *Current Protocols in Pharmacology*, 25, 1–22, 2004.
11. S. Fulda, A. M. Gorman, O. Hori and A. Samali, Cellular stress responses: Cell survival and cell death, *International Journal of Cell Biology*, 2010, 1–23, 2010.
12. M. V. Berridge, P. M. Herst and A. S. Tan, Tetrazolium dyes as tools in cell biology: New insights into their cellular reduction, *Biotechnology Annual Review*, 11, 127–152, 2005.

13. N. Marshall, C. Goodwin and S. Holt, A critical assessment of the use of microculture tetrazolium assays to measure cell growth and function, *Growth Regulation*, 5, 69–84, 1995.

14. T. Mosmann, Rapid colorimetric assay for cellular growth and survival: Application to proliferation and cytotoxicity assays, *Journal of Immunological Methods*, 65, 55–63, 1983.

15. M. V. Berridge and A. S. Tan, Characterization of the cellular reduction of 3-(4, 5-dimethylthiazol-2-yl)-2, 5-diphenyltetrazolium bromide (MTT): Subcellular localization, substrate dependence and involvement of mitochondrial electron transport in MTT reduction, *Archives of Biochemistry and Biophysics*, 303, 474–482, 1993.

16. M. B. Hansen, S. E. Nielsen and K. Berg, Re-examination and further development of a precise and rapid dye method for measuring cell growth/cell kill, *Journal of Immunological Methods*, 119, 203–210, 1989.

17. F. Denizot and R. Lang, Rapid colorimetric assay for cell growth and survival: Modifications to the tetrazolium dye procedure giving improved sensitivity and reliability, *Journal of Immunological Methods*, 89, 271–277, 1986.

18. A. H. Cory, T. C. Owen, J. A. Barltrop and J. G. Cory, Use of an aqueous soluble tetrazolium/formazan assay for cell growth assays in culture, *Cancer Communications*, 3, 207–212, 1991.

19. J. A. Barltrop, T. C. Owen, A. H. Cory and J. G. Cory, 5-(3-carboxymethoxyphenyl)-2-(4,5-dimethylthiazolyl)-3-(4-sulfophenyl)tetrazolium, inner salt (MTS) and related analogs of 3-(4,5-dimethylthiazolyl)-2,5-diphenyl tetrazolium bromide (MTT) reducing to purple water-soluble formazans as cell-viability indicators, *Bioorganic & Medicinal Chemistry Letters*, 1, 611–614, 1991.

20. K. D. Paull, R. H. Shoemaker, M. R. Boyd, J. L. Parsons, P. A. Risbood, W. A. Barbera, M. N. Sharma et al., The synthesis of XTT: A new tetrazolium reagent that is bioreducible to a water-soluble formazan, *Journal of Heterocyclic Chemistry*, 25, 911–914, 1988.

21. L. Weir, D. Robertson, I. M. Leigh and A. A. Panteleyev, The reduction of water-soluble tetrazolium salt reagent on the plasma membrane of epidermal keratinocytes is oxygen dependent, *Analytical Biochemistry*, 414, 31–37, 2011.

22. T. T. Yang, P. Sinai and S. R. Kain, An acid phosphatase assay for quantifying the growth of adherent and nonadherent cells, *Analytical Biochemistry*, 241, 103–108, 1996.

23. V. Knowles and W. Plaxton, Protein extraction, acid phosphatase activity assays and determination of soluble protein concentration, *Bio-Protocol*, 3, e889, 2013.

24. D. Shum, C. Radu, E. Kim, M. Cajuste, Y. Shao, V. E. Seshan and H. Djaballah, A high density assay format for the detection of novel cytotoxic agents in large chemical libraries, *Journal of Enzyme Inhibition and Medicinal Chemistry*, 23, 931–945, 2008.

25. G. Repetto, A. del Peso and J. L. Zurita, Neutral red uptake assay for the estimation of cell viability/cytotoxicity, *Nature Protocols*, 3, 1125–1131, 2008.

26. S. Z. Zhang, M. M. Lipsky, B. F. Trump and I. C. Hsu, Neutral red (NR) assay for cell viability and xenobiotic-induced cytotoxicity in primary cultures of human and rat hepatocytes, *Cell Biology and Toxicology*, 6, 219–234, 1990.

27. A. L. Niles, R. A. Moravec, P. E. Hesselberth, M. A. Scurria, W. J. Daily and T. L. Riss, A homogeneous assay to measure live and dead cells in the same sample by detecting different protease markers, *Analytical Biochemistry*, 366, 197–206, 2007.

28. M. J. Humphries, Cell-substrate adhesion assays, *Current Protocols in Cell Biology*, 9(1), 1–11, 2001.

29. R. Tripathy, A. Ghose, J. Singh, E. R. Bacon, T. S. Angeles, S. X. Yang, M. S. Albom, L. D. Aimone, J. L. Herman and J. P. Mallamo, 1, 2, 3-Thiadiazole substituted pyrazolones as potent KDR/VEGFR-2 kinase inhibitors, *Bioorganic and Medicinal Chemistry Letters*, 17, 1793–1798, 2007.

4 Cell Proliferation Assays

4.1 INTRODUCTION

Cell proliferation or cell division plays a crucial role in tissue and cellular homeostasis for proper growth, development, and maintenance of an organism. In a cell proliferation assay, the variation in the proportion of cells that undergoes cell division or the total number of cells that are dividing in a cell culture is quantified. The parameters that can be assessed in these are cytotoxicity, examination of circumstances of cell activation, effects of different pharmacological agents, or fluctuations in growth factors [1]. The following three classes of cell proliferation assays are mainly used:

- DNA replication rate during cell division can be examined using radioactive or labeled nucleotide analogs like ^{3}H-thymidine and BrdU.
- Tetrazolium dyes, resazurin, and ATP assays can quantify cell proliferation based on metabolic activity.
- Cell proliferation antigen-based approaches can target antigens/markers like topoisomerase IIB, Ki-67, PCNA, and phosphohistone H3 present in proliferating cells.

The choice of method depends partly on the type of cell used in the study and the path preferred for proliferation measurement. For studies measuring metabolic activity in a population of cells either plated or in suspension, tetrazolium salts with colorimetric detection can be adopted as discussed in Chapter 3. Alternatively, if DNA synthesis is the major focus, labeling with BrdU, tagging it with a fluorescently labeled antibody, and then examining single cells by fluorescence-activated cell sorting (flow cytometry) can be used. Irrespective of the type of cell proliferation assay chosen, all these assays are virtually reliable and robust.

4.2 DNA LABELING ASSAYS

Since the DNA has to replicate prior to cell division, the rate of DNA synthesis reflects cell replication. As DNA synthesis is a critical phase involved in cell proliferation, the quantification of DNA synthesis provides very supportive evidence for judging cell multiplication [2]. The measurement of the rate of DNA synthesis in the presence of a label remains one of the most widely used assays that render precise and consistent results. These assays use labeled DNA precursors like ^{3}H-thymidine or bromodeoxyuridine (BrdU), which when added into a cell culture sample, the cells that are going to divide incorporate this labeled nucleotide into their DNA. The integration of the labeled nucleotides is quantified after incubation either by measuring the net amount of cells with labeled DNA in a cell population by liquid

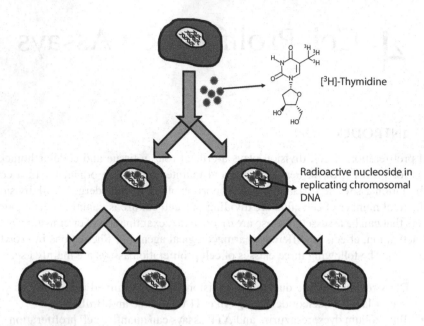

[³H]-Thymidine

Radioactive nucleoside in replicating chromosomal DNA

FIGURE 4.1   ³H-Thymidine incorporating replicating cells.

scintillation counting, or by identifying the labeled nuclei microscopically [3,4]. The inclusion of the labeled precursor into the DNA is related directly to the extent of cell division happening in the cell culture.

4.2.1   [³H]-THYMIDINE ASSAY

Tritiated thymidine is the most widely used radiotracer for imaging tumor proliferation [5,6]. The cells in culture incorporate the radioactive nucleoside ³H-thymidine into their nascent replicating chromosomal DNA during mitotic cell division. The DNA uptake by the proliferating cells is measured after washing and lysing the cells. The radioactivity in DNA extracted from the cells is then estimated employing a scintillation beta-counter to see the extent of cell division that has taken place in response to a test agent (Figure 4.1). The amount of labeled thymidine incorporated during DNA synthesis reflects cell division and hence serves as a marker to measure cell proliferation [7].

Procedure:

1. Seed 1×10^6 cells/mL in autoclaved glass tubes.
2. Incubate with the test drug for the desired period.
3. Add 0.5 µCi/mL of [³H]-thymidine into the culture medium and incubate further for 2 h.
4. Arrange the filters in a Millipore filtration unit and wet with cold distilled water.
5. Filter the contents of the tube and wash the cells three times with 3 mL 10% trichloroacetic acid (TCA).

6. Wash further twice with 3 mL cold methanol and once with 3 mL cold distilled water.
7. Discard the filtrate into a container designed for disposing of radioactive wastes.
8. Remove the filters and transfer them into liquid scintillation vials.
9. Dry the filters at 37°C for 30 min.
10. Add scintillation cocktail into the vials and mix properly.
11. Count in an appropriate beta counter.

Advantages:

- Sensitive and produce accurate data on DNA synthesis.
- Thymidine, being a natural compound, is readily taken up by the cells and incorporated into the DNA.
- It directly measures proliferation.

Limitations:

- Hazards and hassle in using and disposing of radioisotopes cause various concerns.
- Long half-life of ^3H.
- Radioactive waste disposal is costly.
- Requires extensive sample preparation.
- The method is sample destructive.

4.2.2 BrdU Incorporation Assay

The synthetic pyrimidine analog of thymidine, 5-bromo-2′-deoxy-uridine (BrdU) can be employed as a substitute for thymidine and can be integrated into cellular DNA during the synthesis phase of the cell cycle to study proliferation of a given cell population. The incorporation can be detected and quantified using cellular enzyme nonradioactive immunoassay exploiting monoclonal antibodies directed against BrdU [8–10]. This assay incorporates a few additional steps compared to tritiated thymidine incorporation assay because cells have to be incubated with a BrdU-specific monoclonal antibody, followed by a secondary antibody that acts as a reporter (Figure 4.2). The reporter signal can be measured by colorimetric, chemiluminescent, or fluorescent methods. This assay can also be used to detect apoptosis.

Procedure:

1. Culture the cells in an appropriate culture vessel or microtiter plate and incubate with test drugs for the desired duration.
2. Dissolve 100 mg BrdU in 32.5 mL anhydrous DMSO to prepare 10 mM of stock solution.
3. Dilute 10 μL of this stock solution in 10 mL of culture medium to make a 10 μM labeling solution.
4. Remove culture medium from cells and replace with BrdU labeling solution.

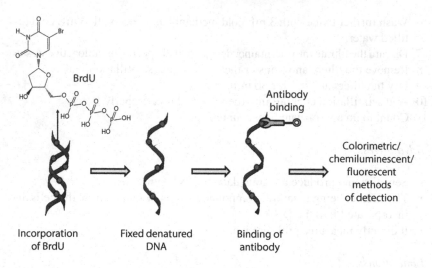

BrdU

Antibody binding

Colorimetric/ chemiluminescent/ fluorescent methods of detection

Incorporation of BrdU

Fixed denatured DNA

Binding of antibody

FIGURE 4.2 BrdU incorporation into replicating cells.

5. Incubate cells at 37°C for 2 h.
6. Remove labeling solution and wash with PBS (3 times, 2 min each).
7. Remove PBS and add 1 mL of 3.7% formaldehyde in PBS to each culture well.
8. Incubate for 15 min at room temperature and wash with PBS (3 times, 2 min each).
9. Remove PBS, add Triton X-100 permeabilization buffer to each well and incubate for 20 min at room temperature.
10. Remove permeabilization buffer, add 1N HCl, and incubate 10 min on ice.
11. Remove this solution, add 2N HCl, and incubate 10 min at room temperature.
12. Add phosphate/citric acid buffer (pH 7.4) and incubate at room temperature for 10 min.
13. Wash using Triton X-100 permeabilization buffer (3 times, 2 min each).
14. Remove this solution and add antibody staining buffer.
15. Add anti-BrdU primary antibody and incubate overnight at room temperature.
16. Wash with Triton X-100 permeabilization buffer (3 times, 2 min each).
17. Add fluorescently labeled secondary antibody and incubate 1 h at room temperature.
18. Add PBS and take images using appropriate filters to quantify by an ELISA plate reader.

Advantages:

- Widely preferred as it is neither radioactive nor myelotoxic at labeling concentrations.
- Transfer/trypsinization of cells is not essential since the assay is carried out in a single microplate.

Limitations:

- BrdU usage is a potential health hazard as it can replace thymidine during DNA replication and cause mutations.
- Relatively laborious process.

4.3 METABOLIC ASSAYS

Metabolic activity of a particular cell population can be employed to measure the rate of cell proliferation.

4.3.1 DYE ASSAYS

The reduction of tetrazolium salts (MTT, XTT, MTS, and WST1) or Alamar blue by metabolically active cells which is discussed in Chapter 3 can also be used as markers for measuring cell proliferation. MTT is mainly an endpoint assay as the formazan crystals produced during reduction must be dissolved in an appropriate solvent. The other dyes can be used to continuously monitor the dynamic changes in cell proliferation over time as they are soluble in culture media and are nontoxic. WST1 is preferred to XTT as it is more sensitive, needs no additional factors, reduces more efficiently, and displays faster color development. Alamar blue is also sensitive and can detect as few as 100 cells in an assay well. All these redox dyes can be quantified with a wide choice of instruments such as standard spectrophotometers or spectrofluorometers or plate readers for conventional or high-throughput studies.

4.3.2 ATP ASSAYS

This approach is well suited to quantify cell proliferation by taking advantage of the tight regulation of intracellular ATP. As soon as the cells lose their membrane integrity, their capability to synthesize ATP vanishes rapidly and the endogenous ATPases instantly use up any remaining ATP from the cytoplasm. Since the dying or dead cells hold little or no ATP, there is a direct connection between cell number and the ATP concentration measured. The bioluminescence-based detection of ATP in either a cell lysate or extract using the enzyme luciferase and its substrate luciferin offers an extremely sensitive readout. Luciferase produces light proportional to the ATP concentration which can be sensed by a luminometer (Figure 4.3) or any microplate reader capable of perceiving luminescent signals.

The ATP detection reagent is a combination of several elements: (1) detergent for cell lysis, (2) ATPase inhibitors for stabilizing ATP discharged from the lysed cells, (3) luciferin substrate, and (4) luciferase for catalyzing the reaction that emits luminescence. The luminescent light signal attains a stable state within 10 min after the reagent addition and glows with a half-life of more than 5 h [11].

Procedure:

1. Culture the cells to make a final volume of 100 μL in each well in white opaque walled microwell assay plates.

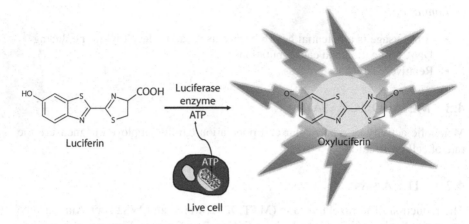

FIGURE 4.3 Bioluminescence-based detection of ATP using the enzyme luciferase which produces light proportional to the ATP concentration.

2. Expose the cells to test drug for the desired period.
3. Equilibrate plates for 30 min to confirm uniform temperature across the plate during the luminescent assay.
4. Add 100 µL ATP detection reagent to each well.
5. Induce cell lysis by mixing the contents of the well for 2 min on an orbital shaker.
6. Maintain at room temperature for 10 min to stabilize the luminescent signal.
7. Record the luminescence at 560 nm.

Advantages:

- Rapid and highly sensitive.
- Less prone to artifacts compared to other cell proliferation assay methods.
- Fewer number of steps, as it does not require an incubation step with a viable cell population to transform the substrate into a colored compound and also eliminates a plate handling step as the cells need not be returned to the incubator to produce the signal.
- Can detect fewer than 10 cells per well.
- Sensitivity is limited by reproducibility in pipetting replicate samples.

Limitation:

- Can generate uneven luminescent signal within standard plates owing to either irregular cell seeding, temperature gradients, or edge effects in multiwall plates.

4.4 CLONOGENIC REPRODUCTIVE ASSAY

Clonogenic or colony formation assay is mainly grounded on the capability of a single cell in the population to undergo sustained unlimited proliferation to grow into a colony of at least 50 cells [12]. The assay mainly determines cell reproductive death after treatment with cytotoxic agents and is used to establish their long-term effect. The number of cell colonies produced can be counted using a microscope, and a comparison of the ability of nontreated cells to grow into colonies to that of treated cells represents the surviving or viable fraction.

Procedure:

1. Seed 5×10^6 cells in cell culture bottle or petri dish or 6 well microtiter plates and allow them to adhere and culture for 48 h.
2. Remove the medium, add test agents, and incubate for a specific time.
3. Trypsinize and count the cells.
4. Adjust the cell population to 5×10^3 cells/mL, seed into culture bottles or petri dishes or six well plates and incubate until colonies form (2–3 weeks).
5. Wash the culture with PBS and add 2–3 mL of 6% glutaraldehyde and 0.5% crystal violet mixture and leave for a minimum of 30 min.
6. Take out glutaraldehyde-crystal violet mixture carefully and rinse with water.
7. Leave the plates with colonies to dry in air at room temperature.
8. Count the colonies with more than 50 cells using a stereo microscope and an automatic colony counter pen.
9. Determine the reproductive integrity to form colonies in terms of plating efficiency (PE) or colony forming efficiency (CFE).

$$\text{CFE (PE)} = \frac{\text{Number of colonies formed}}{\text{Number of cell plates}} \times 100$$

$$\text{Surviving fraction} = \frac{\text{Number of colonies formed}}{\text{Number of cells seeded} \times [(PE)/100]}$$

Advantage:

• Colonies can be counted up to at least 50 weeks after staining.

Limitations:

• Tedious protocol and difficult to screen large numbers of samples.
• If number of cell division is less and later they become quiescent, the number of dividing cells may be underestimated, and the colonies will be too small to be counted.
• Establishing growth curves is time-consuming and labor intensive.
• Incorrect results may be obtained if cells are continuously exposed to a drug.
• Culture conditions affect the drug effects on clonogenic survival.

There are few techniques that study the molecules that control the cell cycle either by quantifying them (e.g., ELISA, Western blots, or immunohistochemistry) or their activity (e.g., CDK kinase assays) which can indirectly assess cell proliferation.

REFERENCES

1. G. M. Cooper and R. E. Hausman, *The Cell*. Sinauer Associates, Sunderland, USA, 2000.
2. D. O. Morgan, *The Cell Cycle: Principles of Control*. New Science Press, London, 2007.
3. Z. Darzynkiewicz, X. Huang and M. Okafuji, Detection of DNA strand breaks by flow and laser scanning cytometry in studies of apoptosis and cell proliferation (DNA replication), *DNA Repair Protocols: Mammalian Systems*, 81–93, 2006.
4. S. Öz, G. Raddatz, M. Rius, N. Blagitko-Dorfs, M. Lübbert, C. Maercker and F. Lyko, Quantitative determination of decitabine incorporation into DNA and its effect on mutation rates in human cancer cells, *Nucleic Acids Research*, 42, e152, 2014.
5. C. Kavitha, M. Nambiar, C. A. Kumar, B. Choudhary, K. Muniyappa, K. S. Rangappa and S. C. Raghavan, Novel derivatives of spirohydantoin induce growth inhibition followed by apoptosis in leukemia cells, *Biochemical Pharmacology*, 77, 348–363, 2009.
6. K. K. Chiruvella, V. Kari, B. Choudhary, M. Nambiar, R. G. Ghanta and S. C. Raghavan, Methyl angolensate, a natural tetranortriterpenoid induces intrinsic apoptotic pathway in leukemic cells, *FEBS Letters*, 582, 4066–4076, 2008.
7. A. Duque and P. Rakic, Different effects of bromodeoxyuridine and [³H] thymidine incorporation into DNA on cell proliferation, position and fate, *The Journal of Neuroscience*, 31, 15205–15217, 2011.
8. M. Crane and S. K. Bhattacharya, The use of bromodeoxyuridine incorporation assays to assess corneal stem cell proliferation, *Corneal Regenerative Medicine: Methods and Protocols*, 65–70, 2013.
9. A. L. Cavanagh, T. Walker, A. Norazit and A. C. Meedeniya, Thymidine analogues for tracking DNA synthesis, *Molecules*, 16, 7980–7993, 2011.
10. H. N. Madhavan, Simple Laboratory methods to measure cell proliferation using DNA synthesis property, *Journal of Stem Cells & Regenerative Medicine*, 3, 12–14, 2007.
11. J. J. Lemasters and C. R. Hackenbrock, Firefly luciferase assay for ATP production by mitochondria, *Methods in Enzymology*, 57, 36–50, 1978.
12. N. A. Franken, H. M. Rodermond, J. Stap, J. Haveman and C. Van Bree, Clonogenic assay of cells *in vitro*, *Nature Protocols*, 1, 2315–2319, 2006.

5 Apoptosis Assays

5.1 INTRODUCTION

Cytotoxicity refers to the cell-killing potential of a test compound independent of the mechanisms of death. There are predominantly two distinct mechanisms of cell death: (1) necrosis, the "accidental" cell death, which takes place when cells are subjected to severe physical or chemical stress and (2) apoptosis, the "programmed" cell death that commonly occurs to remove unwanted cells. There are various pathways of cell death classified according to its morphological appearance [1]. A cell is considered to be dead when any one of the following morphological or molecular conditions is encountered: (1) loss of plasma membrane integrity, (2) complete fragmentation of the cell including its nucleus into discrete apoptotic bodies, and (3) its rubbles engulfed by an adjacent cell [2]. Cell death occurs through a variety of biochemically distinct pathways. A few dominant cell death modes are discussed below and depicted in Figure 5.1.

1. *Autophagy*: The cellular components degrade within the intact dying cell inside the autophagic vacuoles. The morphological features include the formation of vacuoles, the disintegration of cytoplasmic contents, and minor chromatin condensation. Later, these autophagic cells are taken up by phagocytosis process [3,4].
2. *Oncosis*: Features cell death that is characterized by cellular and organelle swelling, followed by membrane breakdown with the eventual expulsion of inflammatory cellular contents. The pre-lethal oncotic pathway leads to necrosis with lysis and spillage of contents before undergoing phagocytosis and finally results in inflammation [5,6].
3. *Pyroptosis*: An integrated pro-inflammatory cell death process initiated by a bacterial infection. This caspase-1 mediated exclusive pathway proceeds through the activation of pro-inflammatory cytokines, subsequent cell lysis, and ultimately discharge of inflammatory cellular substances. Pyroptosis can be confirmed by immunoblotting or ELISA for cleaved caspase-1 in isolated cell lysates, as caspase-1 is not involved in any other modes of cell death [7,8].
4. *Necrosis*: Initiated by the activation of receptors on the cell surface that paves the way to the loss of plasma membrane integrity and an uninhibited discharge of apoptotic proteins into the cell. This triggers an inflammatory response, which in turn inhibits the phagosomes from engulfing the dead cells and eventually leads to accumulation of dead tissue [9,10]. Necrosis is an irreversible process which is detrimental to a cell and organism.

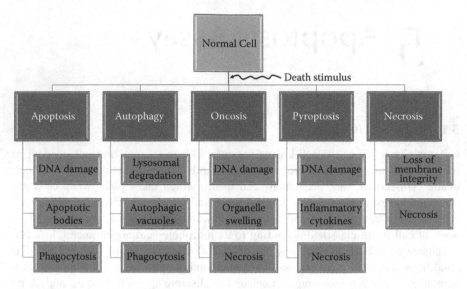

FIGURE 5.1 Different pathways that lead to cell death.

5. *Apoptosis*: The "classical" form of programmed cell death that plays a significant role in the proper maintenance of biological systems. The biochemical and morphological features of apoptotic cell death involve activation of initiator and effector caspases, exposure of phosphatidyl serine on to the cell surface, loss of mitochondrial integrity, DNA fragmentation, chromatin condensation, cellular shrinkage, membrane blebbing and development of apoptotic bodies, followed by phagocytosis by neighboring cells. The apoptotic bodies may undergo lysis and eventually secondary or apoptotic necrosis. It is a highly regulated and organized dynamic process that leads to the breakdown of a cell evading an inflammatory reaction unlike necrosis [11,12].

6. *Anoikis*: Features a specific form of cell death induction with the molecular mechanisms resembling those activated during classical apoptosis process. Anoikis associated cell death is a consequence of a cell being detached from the extracellular matrix [13–15].

Most of the marketed cytotoxic anticancer drugs induce cell death in their susceptible cells through tightly regulated apoptotic cascade process, thereby opening several options to measure the activity of these regulators as well as their functional implications. Numerous assays have been developed for detecting and quantifying apoptotic cells [16]. It is often advantageous to use a multi-parametric approach when studying apoptotic events. As certain characteristics of apoptosis might partly overlap with necrosis, while some other features might appear only transiently, it is highly essential to use two or more discrete assays to establish apoptosis during a cell death event. Familiarity on the pros and cons of every assay is vital while choosing the suitable and appropriate detection methods for confirming apoptosis in a sample of cells, tissues, or organs.

There are various diverse assays to measure different features of apoptosis in cell culture, cell lysates, or tissue biopsies that include mainly caspase activation, DNA fragmentation, alterations of the plasma membrane, and mitochondria.

5.2 LIGHT MICROSCOPY

Light microscopy aids in detecting different morphological alterations that occur in the early stages of apoptosis of a cell. Cell shrinkage and round/oval mass of apoptotic cells with eosin stained dark cytoplasm and dense purple chromatin fragments in the nucleus can be observed under a light microscope on hematoxylin/eosin cell staining [17].

Procedure:

1. Seed cells in a petri dish and incubate with test compounds for desired time.
2. Wash with PBS and stain rehydrated cells in hematoxylin solution for 20–40 min.
3. Wash with tap water for 1–5 min, until the culture turns blue.
4. Separate the cells in 70% ethanol containing 1% HCl for 5 sec to eliminate excess dye, letting nuclear details develop.
5. Wash 1–5 min in tap water until blue.
6. Stain in eosin solution for 10 min and wash in tap water for 1–5 min.
7. Dehydrate, mount, and observe the cells.

Advantage:

• Fairly reliable and inexpensive method.

Limitations:

• Quantitative assessments do not yield reproducible results.
• Less sensitive, time-consuming, and susceptible to many errors.
• Lens of high power should be used as fewer apoptotic cells are detected at low magnification.
• The risk of interobserver variability during result analysis.
• The number of apoptotic cells scored is usually less as light microscopic detection relies solely on cell morphology.

5.3 WESTERN BLOT CASPASE ACTIVATION ASSAY

Activation of caspase enzymes, which aids in protein cleavage eventually leading to cellular disassembly, is a unique feature observed in the initial phases of apoptosis. Caspase-3 is an executive molecule that normally exists in its proenzyme state in the cellular cytoplasm through which apoptosis signals normally transduce. Caspase-3 that gets activated during the initial stage of apoptosis cleaves its target substrates at aspartic acid residues in a well-defined sequence [18,19]. As caspase-3 activity declines significantly toward the later stages of apoptosis, *in situ* detection of activated caspase-3 can be a more distinct, sensitive, and direct indicator of early apoptosis [20].

Western blotting is an important technique used to identify specific proteins from a complex mixture of proteins. In this technique, the underlying principle is antigen-antibody interaction followed by immune detection. The procedure is accomplished by three steps: (1) separation by size or molecular weight by gel electrophoresis, (2) protein transfer from the gel to a solid support, and (3) identification of target protein by using a suitable primary and secondary antibody. As the primary antibodies bind to the protein of interest, only a band will be visible, and its thickness can be used as an indication of protein concentration in comparison to standard protein.

Procedure:

1. Harvest cultured cells with lysis buffer (containing Tris-HCl, pH 8, NaCl, Triton X-100, sodium deoxycholate, sodium dodecyl sulfate, NaF, and protease inhibitors) after drug exposure of the desired duration.
2. Transfer the cell suspension into a cooled microfuge tube and agitate for 30 min at 4°C.
3. Spin at 16,000 × g for 20 min at 4°C and transfer the supernatant to a fresh tube.
4. Extract, purify, and quantify the total protein concentration.
5. Isolate target protein by SDS-PAGE (protein is denatured, and an overall negative charge is imparted).
6. Transfer proteins to the nitrocellulose or PVDF membranes.
7. Block nonspecific antibody sites by incubating in PBST buffer (PBS with Tween 20) for 20 min to 1 h at room temperature.
8. Incubate with ~1:1000 to 1:2000 of caspase-3 monoclonal antibody (MAb) or polyclonal antibody (PAb) at room temperature for 1~2 h or at 4°C overnight.
9. Remove the nonspecifically bound antibody from the membrane by washing with PBST for 3 × 10 min.
10. Incubate with ~1:5000 to 1:10000 of antimouse IgG secondary antibody conjugated with horseradish peroxidase (HRP) or alkaline phosphatase (AP) at room temperature for 30 min to 1 h.
11. Wash the membrane using PBST for 4 × 15 min.
12. Develop with ECL (enhanced chemiluminescence) or BCIP/NBT (5-bromo-4-chloro-3-indolyl phosphate/nitro blue tetrazolium).
13. Record the findings by autoradiography, photographing, or fluorescence detection.

Advantages:

• Highly sensitive and reliable method.
• Can detect even low quantity of a specific protein.

Limitations:

• Can display unusual or unexpected bands.
• Sometimes bands may not show up.

- Bands can be faint or weak.
- High background display on the blot.
- Uneven or patchy spots on the blot.

5.4 ANNEXIN-V STAINING

Another distinguishing feature of apoptotic cell death is the translocation of phosphatidylserine (PS) that is typically restricted to the inner leaflet of plasma membrane to the exterior of cell membrane in response to pro-apoptosis stimuli. This flip over mechanism is caspase dependent and embodies an "eat-me" signal for the engulfment by phagocytic cells resulting in binding to a recognized PS receptor [21]. Exposure of PS thus enables the removal of apoptotic cells without any release of cellular contents and also not provoking any inflammatory reaction.

Annexin-V, a 35 to 36 kilodalton Ca^{2+} dependent phospholipid-binding protein that has great affinity for the anionic phospholipid PS can be used to detect exposed PS in apoptotic cells (Figure 5.2) [22]. Annexin-V staining is among the most sensitive methods to identify ongoing apoptosis and several Annexin-V derivatives attached with different fluorochromes are available that provide versatile options to confirm early and late apoptosis by either multicolor flow cytometry or fluorescence microscopy [23].

As the necrotic cells are also stained by Annexin-V on plasma membrane rupture, double-staining using membrane-impermeable DNA dyes like PI can be adopted to differentiate cell death due to necrosis and apoptosis. In these double-staining assays, viable cells are doubly negative to both Annexin-V and PI, while the early apoptotic cells are Annexin-V-positive but PI-negative, and the cells that undergo necrosis are doubly positive to both Annexin-V and PI (Figure 5.3).

Procedure:

1. Seed 1×10^6 cells and incubate with the test agent for the desired duration.
2. Wash cells with PBS and centrifuge ($200 \times g$) for 5 min at 15–25°C.
3. Incubate the cells with 100 µL of Annexin-V reagent (containing 10 µL of 10X binding buffer, 10 µL PI, 1 µL Annexin-V-FITC, 79 µL dH₂O) for 10–15 min at 15–25°C in the dark.

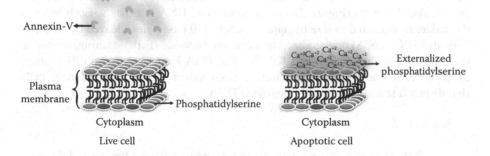

FIGURE 5.2   Schematic representation of the Annexin-V assay.

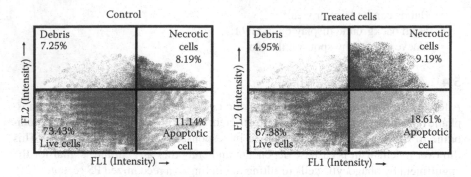

FIGURE 5.3   Annexin-V-PI staining to differentiate live, apoptotic, and necrotic cells.

4. Add 100 µL binding buffer (containing 10 mL of 1M HEPES/NaOH, pH 7.4, 30 mL of 5 M NaCl, 5 mL of 1 M KCl, 1 mL of 1 M MgCl$_2$, 1.8 mL of 1 M CaCl$_2$, 52.2 mL dH$_2$O) to the stained cells.
5. Analyze by flow cytometer/fluorescence microscopy.

Advantages:

- Sensitive and rapid method.
- Can confirm the activity of initiator caspases.

Limitation:

- Annexin-V, when used alone, cannot differentiate necrosis from apoptosis. Hence, used in combination with PI that stains necrotic cells.

5.5   DNA LADDER ASSAY

The DNA laddering technique allows visualizing the DNA fragmentation that occurs due to endonuclease cleavage during apoptosis. DNA is extracted from the drug-treated lysed apoptotic cell homogenate and is precipitated using agarose or polyacrylamide or polyethylene glycol (PEG). The DNA fragments that remain in the supernatant can be exposed to agarose gel electrophoresis to exhibit a characteristic "ladder" pattern (Figure 5.4) of discontinuous DNA fragments. Each band in the ladder is separated in size by approximately 180 base pairs, cleaved by caspase-activated DNase (CAD), visualized by ethidium bromide (EtBr) staining, or can be quantified using fluorescent dyes [24,25]. The DNA ladder assay can differentiate apoptotic cells that reveal a characteristic DNA ladder pattern, from necrotic cells that display a smear of randomly degraded DNA.

Procedure:

1. Culture the cells, expose the cells to test compounds for the desired duration and add 1 mL of trypsin to cell monolayer in 100 mm dishes.

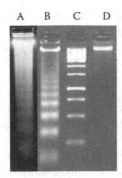

FIGURE 5.4 Agarose gel showing DNA bands stained with Ethidium bromide: Lane (A) showing necrosis; Lane (B) showing a typical ladder formation; Lane (C) showing ladder pattern of 1 kb DNA marker; and Lane (D) control DNA are depicted.

2. Harvest cells by centrifugation at 2500 rpm for 5 min and wash cell pellets once with PBS.
3. Add 100 μL of lysis buffer (1% NP-40 in 20 mM EDTA, 50 mM Tris-HCl, pH 7.5) and keep for 10 sec.
4. Centrifuge at 3000 rpm for 5 min and collect the supernatant.
5. Treat the aqueous phase with 40 mg/mL RNase for 2 h at room temperature.
6. Add an equal volume of phenol/chloroform and extract the genomic DNA with chloroform followed by ethanol precipitation in the presence of 0.3 M sodium acetate.
7. Centrifuge at 10,000 rpm for 5 min at 4°C to obtain the DNA pellet.
8. Resuspend the DNA pellet in 100 μL Tris-EDTA buffer.
9. Introduce 10 μg of DNA to 1.5% agarose gel containing EtBr.
10. Carry out electrophoresis with 1 × TAE (Tris base + acetic acid + EDTA) buffer at a constant voltage of 50 V for 1 h.
11. Visualize the bands under a UV transilluminator at 365 nm.

Advantages:

• Quick screening of apoptotic changes in cell populations.
• Does not require any special laboratory equipment as this method allows working with cell lysates.
• Simple with high sensitivity and easy to perform.

Limitations:

• Time-consuming.
• Not advised in instances with low numbers of apoptotic cells.
• As DNA fragmentation happens toward the later stages of apoptosis, the absence of a DNA ladder does not exclude the possibility that cells are going through early apoptosis.

- DNA fragmentation can also occur while preparing, creating difficulty to generate a nucleosome ladder.
- Necrotic cells can also sometimes produce DNA fragments.

5.6 ELISA

The isolation and electrophoretic examination of DNA is a tedious and time-consuming process. An alternate and very elegant method for measuring apoptotic cells is the immunological detection of histone-complexed low molecular weight DNA fragments by an enzyme-linked immunosorbent assay (ELISA). This sensitive nonradioactive immunoassay reliably detects mono- and oligo-nucleosomes in the cytoplasm that are produced by internucleosomal DNA cleavage following the disintegration of the nuclear membrane in the course of apoptosis [26]. The essential components of antibody-based immunoassay systems to measure cytoplasmic nucleosomes are threefold: an antigen to detect and quantitate; a specific antibody to identify this antigen; and a system to measure the amount of antigen in a given sample.

ELISA serologies are done in multiwall microtiter plates so that the dilutions of serum can be easily prepared and tested. The wells are coated with antigen of interest and are filled with dilutions of serum. If antibody against the antigen is present, it will bind to the antigen fixed to the bottom of the well. Only antigen-specific antibodies will bind to the well. The wells are washed to remove unbound antibodies. Next, a solution of secondary antibodies is added which is covalently conjugated to an enzyme. The wells are washed to remove unbound enzyme conjugated antibody. Finally, a solution of chlorogenic enzyme substrate is added. The interaction of substrate with the enzyme on the secondary antibody generates visible color which can be quantified with an electronic plate reader that is proportional to the amount of apoptotic cells (Figure 5.5).

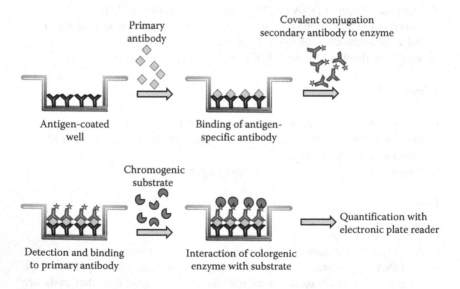

FIGURE 5.5   ELISA for apoptotic cell quantification.

Procedure:

1. Culture 1×10^4 cells in a microwell plate and treat them with test agents for the desired period.
2. Centrifuge microplate (200 × g) for 10 min at 15–25°C to obtain cell pellets and discard the supernatant.
3. Incubate treated cells for 30 min at 15–25°C with lysis buffer.
4. Repeat the microplate centrifugation (200 × g) for 10 min at 15–25°C.
5. Transfer the supernatant aliquot to streptavidin-coated microplate.
6. Incubate the above lysate with immunoreagent comprising anti-histone (biotin labeled) and anti-DNA (peroxidase conjugated) for 2 h at 15–25°C. Antibody-nucleosome complexes are bound to the microplate by the streptavidin.
7. Wash the microplate wells containing immobilized antibody-histone complexes three times with incubation buffer at 15–25°C to remove nonimmunoreactive cell components.
8. Add chromogenic substrate solution (peroxidase substrate ABTS) to the wells and incubate for 15 min at 15–25°C.
9. Measure the absorbance of immobilized antibody-histone complexes at 405 nm.

Advantages:

• Specific, sensitive, and convenient method.
• Quick results.
• As capture antibody only holds the specific antigen, antigens of very low or unknown concentration can be sensed and detected.
• Radioisotopes or a costly radiation counter not required.

Limitations:

• Only monoclonal antibodies can be employed as matched pairs.
• Monoclonal antibodies can be expensive.
• Required types of monoclonal antibodies may not be available.
• Negative controls can give positive results in cases where blocking solution is ineffective as secondary antibody or antigen of unknown sample can bind to the open sites in the well.
• Microwells must be read immediately as enzyme/substrate response is short term.

5.7 TUNEL ASSAY

Terminal deoxynucleotidyl transferase nick end labeling (TUNEL) is designed to identify apoptotic cells that endure extensive degradation of DNA during the late phase of apoptosis. The enzyme deoxynucleotidyl transferase (TdT) catalyzes the addition of deoxyuridine triphosphate (dUTP) to the blunt ends of DNA

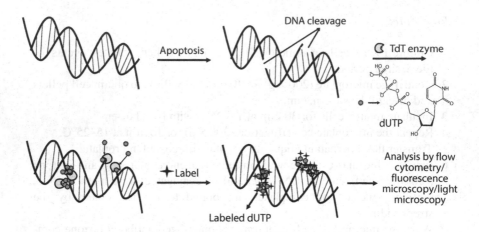

FIGURE 5.6   TUNEL assay to detect apoptosis.

double-strand breaks independent of a template. TUNEL assay relies on the ability of TdT to identify the nicks, which are a few nucleotides of a DNA sequence that is replaced with their labeled analogs and add labeled UTP to the 3′-end of the DNA fragments [27,28]. The dUTP can then be tagged with selected probes to allow detection by flow cytometry, fluorescence microscopy, or light microscopy (Figure 5.6).

Procedure:

1. Culture the desired number of cells and collect them using centrifugation.
2. Wash cells using PBS and suspend $1–2 \times 10^7$/mL cells in PBS.
3. Transfer 100 µL of cell suspension to a V-bottomed 96-well plate.
4. Incubate with test compounds for the desired duration.
5. Wash cells with PBS, add 100 µL of 2% formaldehyde in PBS to fix the cells at pH 7.4, and incubate on ice for around 15 min.
6. Centrifuge to collect the cells; wash once with 200 µL PBS and further postfix with 200 µL of 70% ice-cold ethanol.
7. Collect cells by centrifugation and wash twice with 200 µL PBS.
8. Suspend $1–5 \times 10^5$ cells in 50 µL of TdT equilibration buffer (2.5 mM Tris-HCl [pH 6.6], 0.2 M potassium cacodylate, 2.5 mM $CoCl_2$, 0.25 mg/mL BSA).
9. Incubate the cell suspension at 37°C for 10 min with mild mixing.
10. Resuspend cells in 50 µL of TdT reaction buffer (0.5 U/µL of TdT and 40 pmol/µL biotinylated dUTP).
11. Incubate the cell suspension at 37°C for 30 min with intermittent gentle mixing.
12. Collect cells by centrifugation and wash with 200 µL PBS.
13. Resuspend the cells in 100 µL TdT staining buffer (4× saline sodium citrate—0.6 M NaCl, 60 mM sodium citrate, 2.5 µg/mL fluorescein isothiocyanate-conjugated avidin, 0.1% Triton X-100, and 1% BSA).

14. Incubate the cell suspension at room temperature in the dark for 30 min.
15. Centrifuge, collect the cells, and wash twice with 200 μL PBS.
16. Resuspend $2-8 \times 10^6$/mL cells in PBS.
17. Observe the cells using fluorescence/confocal microscopy or flow cytometry.

Advantages:

- Highly sensitive assay that can detect a cell via fluorescence microscopy or ~100 cells through flow cytometry.
- Fast, as it can be completed within 3 h.
- Highly reproducible method, yielding precise results.
- Loss of small fragments of DNA can be avoided by paraformaldehyde fixation before permeabilization.
- Direct labeling of the 3′-hydroxyl terminal of DNA breaks aids in measuring the lesions which are perceptible at the molecular level.
- Detects very early events of DNA breaks in apoptosis, which are difficult to identify based on morphological changes.
- Besides DNA breaks, DNA content can also be measured.

Limitations:

- Expensive.
- The number of strand breaks necessary for detection is not known.
- Necrotic cells can produce false positive results.
- The detergent used to permeabilize the cells is extremely fragile and cells can be lysed when pipetted or vortexed.
- The specificity and sensitivity of the process rely on the fixative employed, pretreatment, and concentration of TdT enzyme.

5.8 COMET ASSAY

Comet assay or single-cell gel electrophoresis is a simple, standard, and convenient technique that measures DNA strand breaks in eukaryotic cells [29]. Cells are encapsulated in a low-melting-agarose suspension and lysed in neutral or alkaline (pH > 13) conditions with a detergent/high salt. The supercoiled broken fragments of damaged DNA linked to the nuclear matrix lose their supercoiling and become free to extend and migrate toward the positive pole (anode) in the gel [30]. Electrophoresis of suspended lysed cells results in a DNA migration pattern resembling comets. The extent of DNA damage can be visually analyzed using fluorescence microscopy and measured by manual scoring or automatic imaging software. The length and the intensity of the tail can be measured; the tail length is proportional to the damage, and the intensity of the comet tail relative to the head reflects the number of DNA breaks. Apoptotic cells appear in the form of comet-like structures with long tails and small heads, whereas necrotic cells present big nuclear remnants with virtually invisible tails. Live cells exhibit a large head with only a small/no tail (Figure 5.7).

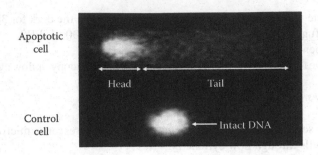

FIGURE 5.7   Picture showing the tailing of denatured DNA.

Procedure:

1. Culture the cells and treat about 1×10^6 cells with test compounds for the desired duration.
2. Remove media and wash with sterile PBS. Trypsinize the cells.
3. Count the cells and mix cells in 0.75% of LMA (low melting agarose) prepared in sterile PBS.
4. Immobilize on Comet slides by layering 200 µL of cell suspension in LMA onto the labeled slides precoated with agarose and immediately place coverslip and keep on ice packs until the agarose film hardens (2–3 min).
5. Remove coverslip and prepare third layer of agarose (200 µL LMA) similarly.
6. Lyse the cells to remove membranes and DNA associated proteins using lysis buffer (2.5 M NaCl, 100 mM EDTA, 10 mM Tris Base, 1% sodium lauryl sarcosinate, and 1% Triton X-100).
7. Unwind and denature DNA by alkaline treatment, pH > 13 (0.6 g NaOH pellets, 250 µL of 200 mM EDTA, pH 10, 49.75 mL dH₂O).
8. Perform electrophoresis at 300 mA, 16 V for 20 min.
9. Remove the slides from the electrophoresis unit and rinse with neutralization buffer.
10. Visualize the damaged, unwound, relaxed DNA that migrates out of the cell by staining.

Advantages:

- Cheap, simple, and quick method.
- Easy quantitative analysis.
- Suitable for many cell types.
- Avoids *in vitro* cultivation step.
- Accurate determination of cell death and DNA damage.
- More sensitive than DNA ladder assay.
- Provides more precise evidence on the magnitude and heterogeneity of DNA damage paralleled to TUNEL assay.

Limitations:

- The DNA damage detected does not relate to fixed mutations.
- An internal reference is required to circumvent experimental variation during electrophoresis.
- Tedious technique attributed to multiple steps and hence requires more time.
- The procedure may impair the cell membrane altering the distribution of apoptotic, necrotic, and/or viable cells.
- Offers qualitative rather than quantitative results.

5.9   ACRIDINE ORANGE/ETHIDIUM BROMIDE STAINING

Acridine orange/ethidium bromide (AO/EB) staining is a simple staining assay designed to view nuclear changes and apoptotic body formation under a fluorescence microscope and quantify apoptosis [31]. An AO-EB combination is recommended as AO stains both live and dead cells, whereas EB stains only those cells that have lost membrane integrity (apoptotic cells). Live cells appear uniformly green under a fluorescent microscope. Cells that are in early stage of apoptosis stain green with bright green dots in the nuclei due to chromatin condensation and nuclear fragmentation, whereas the late apoptotic cells with condensed and often fragmented nuclei incorporate EB and stain orange [32]. In contrast to late apoptotic cells, necrotic cells stain orange, retaining nuclear morphology resembling that of viable cells, with no condensed chromatin (**Figure 5.8**).

Procedure:

1. Culture the cells at $0.5–2.0 \times 10^6$ cells/mL and incubate with test compounds for required duration.
2. Harvest the cells by trypsinization and collect the cell pellet by centrifugation.
3. Wash cells and suspended in cold PBS.
4. Stain with 1 μL of AO/EB solution, mix gently just before microscopy and quantification process, and immediately evaluate the sample.

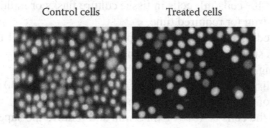

Control cells   Treated cells

FIGURE 5.8   Representation of cell staining using AO-EB staining and their combination image under a confocal microscope.

5. Place 10 µL of cell suspension onto a microscopic slide, cover with a glass coverslip, and observe at least 300 cells in a fluorescence microscope using a 60X objective and a fluorescein filter.

Advantages:

- Rapid and inexpensive method.
- Requires only a small portion of cells from a culture population.

Limitations:

- Manual counting of the cells is tedious.
- Can perform only a few tests at the same time.
- AO is toxic and mutagenic and quenches quickly under standard conditions.

5.10 MITOCHONDRIAL MEMBRANE POTENTIAL ASSAY

Apoptosis-specific mitochondrial changes are difficult to detect. The mitochondrial pathway of apoptosis involves a series of events; (1) permeability of the outer membrane of mitochondria initiated by pro-apoptotic members of the Bcl-2 family, (2) mitochondrial swelling, (3) disruption of the electrochemical gradient across the mitochondrial transmembrane, (4) changes in electron transport, (5) loss of the inner mitochondrial transmembrane potential ($\Delta\Psi$m) due to the electrochemical proton gradient created by the respiratory chain (mitochondrial permeability transition), (6) outer mitochondrial membrane rupture, and (7) eventually release of cytochrome-c and other intermembrane proteins into the cytoplasm [33], which can be studied using Western blot of nuclear, mitochondrial, and cytosolic fractions or by immunocyto-chemistry. Loss of mitochondrial transmembrane potential is a critical parameter that reflects mitochondrial function and can be used as an indicator of mitochondrial damage that occurs during late apoptosis and in necrosis too. The value of $\Delta\Psi$m can be measured employing a range of potentiometric dyes, together with fluorescence microscopy or flow cytometry.

Procedure:

1. Culture 1×10^6 cells/mL cells in tissue culture flasks or multiwell plates.
2. Treat with drug for required time.
3. Remove media containing test compounds, trypsinize and transfer 1 mL each of cell suspension into 12×75 mm polystyrene tubes.
4. Add staining dye solution (JC-1 or Rhodamine 123).
5. Pipette to disperse any cell clumping and incubate for 15–30 min at 37°C.
6. Add 2 mL of assay buffer to each tube and mix gently.
7. Centrifuge the cells at $400 \times g$ for 5 min and remove the supernatant.
8. Suspend the cells in 300–500 µL assay buffer and pipette to remove any cell clumping.
9. Analyze the samples using flow cytometry within 1.5 h.

Advantages:

- Reliable and quick method.
- Staining needs less time.

5.11 CHROMATIN CONDENSATION ASSAY

Cellular morphology is another major parameter that can be considered to ensure apoptotic mode of cell death. Morphologic criteria that define apoptotic cell death include chromatin condensation paralleled by DNA fragmentation with clumping along the nuclear envelope, plasma membrane blebbing, and ultimately formation of small, apoptotic bodies.

During early stages of apoptosis, there is a rapid degradation of nuclease-hypersensitive euchromatin that contains hyperacetylated histones. This feature coincides with the loss of nuclear integrity due to degradation of lamins and reorganization of intra-nuclear protein matrix. These events lead to collapse of nucleus and aggregation of heterochromatin to produce the appearance of condensed apoptotic chromatin. The heterochromatin aggregate is then digested by nucleases to produce oligonucleosomal DNA ladder, which is the hallmark of late apoptosis.

Hoechst staining can be used to determine the chromatin condensation and hence to identify apoptotic cells. Cells are grown in 96-well plates, subjected to test compound treatment, and incubated for 5 min with Hoechst dye which binds at adenine-thymine rich regions of DNA. The dye emits blue fluorescence when excited by UV light at 350 nm when viewed under a fluorescence microscope (Figure 5.9) [34–36].

Procedure:

1. Culture cells in 96-well plates and subject to the test compound treatment for desired period. Arrange a negative control in the absence of test agent.
2. Harvest the cells, wash with cold PBS, and adjust the cell density to ~1 × 10^6 cells/mL in PBS.
3. Prepare a Hoechst 33342 stock solution by dissolving 100 mg of Hoechst dye in 10 mL dH$_2$O to create 10 mg/mL (16.23 mM) solution followed by sonication. Dilute the stock solution 1:2000 in PBS for staining.

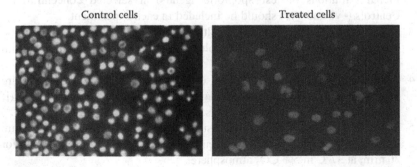
Control cells   Treated cells

FIGURE 5.9 Fluorescence microscopic examination of apoptotic cells stained with Hoechst 33342.

4. Remove about 50 µL of the medium from the wells and add the same amount of staining dye solution.
5. Incubate the cells for 5–10 min on ice in the dark.
6. Remove the medium, wash and view under fluorescence microscope immediately.

Advantages:

- Lesser reaction volumes used in microplates leads to lower costs per assay.
- Better than the conventional tube-based fluorescence assays.

Limitation:

- As Hoechst stain binds to DNA and interferes with DNA replication during cell division, it is potentially mutagenic and carcinogenic. Hence, handling and disposal should be done carefully.

5.12 APOPERCENTAGE™ APOPTOTIC ASSAY

The onset of the execution phase of apoptosis has been linked to translocation of phosphatidylserine (PS) from the interior to the exterior surface of the cell membrane. PS transmembrane movement results in the uptake of the APOPercentage dye by apoptotic committed cells, which continues until blebbing occurs. The dye is selectively incorporated by apoptotic cells, staining them intensely purple-red, which was detected under a conventional microscope and the concentration of the accumulated dye within labeled cells released into solution is measured colorimetrically. Necrotic cells cannot retain the dye and therefore are unlabeled and remain pink [36].

Procedure:

1. Seed $2–5 \times 10^4$ cells/well in 200 µL appropriate culture medium in 96-well plates and incubate the cells at 37°C in 5% CO_2 atmosphere until confluence is reached (~24 h).
2. Prepare dilutions of test apoptotic agent(s) at selected concentrations. Controls (−ve and +ve) should be included in each experiment.
3. Make up double the quantity of culture medium/serum and test apoptotic agent (100 µL/well). Use half the volume to prepare culture medium/serum and 5% v/v dye (100 µL/well).
4. Remove the culture medium from each well and add 100 µL of culture medium/serum and apoptotic test agent, supplemented v/v with serum (if required by the cells) to all the wells.
5. Incubate the cells for desired exposure of drugs and remove the medium. Immediately replace with 100 µL culture medium/serum and incubate for 30 min, at 37°C in 5% CO_2 atmosphere.
   Remove the medium from each well and gently wash the cells twice with 200 µL/well PBS to remove non-cell bound dye.

6. Add sufficient PBS to cover the cells and view immediately with an inverted microscope fitted with a digital camera or add 200 μL dye release reagent, shake the plates for 10 min and read the absorbance at 550 nm.

Advantages:

- Apoptotic cell death can be confirmed as the necrotic cells remain unstained.
- Easy to perform.

5.13 PARP CLEAVAGE ASSAY

Poly (ADP-ribose) polymerase (PARP) colorimetric assay is an alternative to the TUNEL assay. PARP is a protein family that plays a significant role in some cellular processes including DNA replication, cell proliferation and differentiation, DNA repair, and apoptosis [37]. PARP assay analyzes the PARP activity in cells and tissues by identifying its integration onto the histone proteins. PARP cleavage is recognized as one of the established features of apoptosis wherein caspase-3 and -7 split PARP to produce two fragments—85 kDa and 25 kDa, which can be analyzed by Western blot of PARP.

Procedure:

1. Harvest the cultured cells and incubate with test drugs for desired exposure.
2. Remove media and incubate with lysis buffer.
3. Quantify total protein concentration.
4. Isolate the target protein by SDS-PAGE and transfer them to PVDF or nitrocellulose membranes.
5. Incubate in PBST or TBST with 1% bovine serum albumin or 1% non-fat milk blocking buffer, for 20 min to 1 h at 37°C.
6. Treat with 1:1000~1:2000 of either PARP1 monoclonal antibody or PARP1 polyclonal antibody at room temperature and incubate for 1~2 h or at 4°C overnight.
7. Wash the membrane using TBST or PBST for 3 × 10 min.
8. Incubate with 1:5000~1:10,000 of antihuman IgG conjugated with horseradish peroxidase (HRP) or alkaline phosphatase (AP) at room temperature for 30 min to 1 h.
9. Wash the membrane using TBST or PBST for 4 × 15 min.
10. Develop using NBT/BCIP or ECL.

5.14 FLOW CYTOMETRY

Flow cytometry can differentiate apoptotic from nonapoptotic cells using DNA staining. It is a method of choice that allows analysis of multiple parameters simultaneously, pertaining to both physical and chemical features of single cells that flow through an optical apparatus enabling accurate measurement of apoptosis. The plasma membrane of apoptotic cells eliminates the uptake of DNA-specific fluorochromes like Trypan blue, propidium iodide, DAPI (4′,6-diamidino-2-phenyl-indole), acridine orange, and Hoechst dyes. The apoptotic cells can be stained using a fluorescent dye and are passed through a single wavelength light beam, wherein each cell that passes through this scatters light to a certain degree. This forward scatter against side scatter differentiates apoptotic cells in a suspension from others and permits identification of the immune phenotype of the cells going through apoptosis [38]. The results of flow cytometric studies are represented in terms of histograms depicting the percentage of cells in different stages of the cell cycle. In the DNA histograms, the "sub-G1" peak accounts for the apoptotic cells with degraded DNA containing hypodiploid DNA content [39,40].

The integrity of cellular plasma membrane is preserved during early apoptosis and hence certain dyes like PI do not enter the cell. In contrast, apoptotic cells appear brighter than the controls on exposure to specific dyes like Hoechst 33342. Thus, simultaneous staining with dyes belonging to these two categories enables indentifying, discriminating, and quantifying apoptotic cells from viable as well as necrotic cells based on the light scattering principle.

Advantages:

• Easy, quick, and accurate quantification of apoptosis in viable as well as fixed single cells.
• Reflects the direct connection between apoptosis initiation by test agents and their cell cycle phase specificity.

Limitations:

• Multiple steps in the protocol are time-consuming.
• Require enzyme pretreatment to liberate the individual cells for analysis.

REFERENCES

1. G. Kroemer, L. Galluzzi, P. Vandenabeele, J. Abrams, E. Alnemri, E. Baehrecke, M. V. Blagosklonny et al., Classification of cell death: Recommendations of the Nomenclature Committee on Cell Death 2009, *Cell Death & Differentiation*, 16, 3–11, 2009.
2. L. Galluzzi, M. Maiuri, I. Vitale, H. Zischka, M. Castedo, L. Zitvogel and G. Kroemer, Cell death modalities: Classification and pathophysiological implications, *Cell Death & Differentiation*, 14, 1237–1242, 2007.
3. E. H. Baehrecke, Autophagy: Dual roles in life and death? *Nature Reviews Molecular Cell Biology*, 6, 505–510, 2005.
4. Y. Liu and B. Levine, Autopsis and autophagic cell death: The dark side of autophagy, *Cell Death & Differentiation*, 22, 367–376, 2015.
5. P. Weerasinghe and L. M. Buja, Oncosis: An important non-apoptotic mode of cell death, *Experimental and Molecular Pathology*, 93, 302–308, 2012.
6. S. Van Cruchten and W. Van Den Broeck, Morphological and biochemical aspects of apoptosis, oncosis and necrosis, *Anatomia, Histologia, Embryologia*, 31, 214–223, 2002.
7. T. Bergsbaken, S. L. Fink and B. T. Cookson, Pyroptosis: Host cell death and inflammation, *Nature Reviews Microbiology*, 7, 99–109, 2009.

8. I. Jorgensen and E. A. Miao, Pyroptotic cell death defends against intracellular pathogens, *Immunological Reviews*, 265, 130–142, 2015.

9. U. Ziegler and P. Groscurth, Morphological features of cell death, *Physiology*, 19, 124–128, 2004.

10. S. Rello, J. Stockert, V. l. Moreno, A. Gamez, M. Pacheco, A. Juarranz, M. Cañete and A. Villanueva, Morphological criteria to distinguish cell death induced by apoptotic and necrotic treatments, *Apoptosis*, 10, 201–208, 2005.

11. S. Elmore, Apoptosis: A review of programmed cell death, *Toxicologic Pathology*, 35, 495–516, 2007.

12. D. J. Taatjes, B. E. Sobel and R. C. Budd, Morphological and cytochemical determination of cell death by apoptosis, *Histochemistry and Cell Biology*, 129, 33–43, 2008.

13. A. Gilmore, Anoikis, *Cell Death & Differentiation*, 12, 1473–1477, 2005.

14. J. Grossmann, Molecular mechanisms of "detachment-induced apoptosis-Anoikis," *Apoptosis*, 7, 247–260, 2002.

15. S. L. Flink and B. T. Cookson, Apoptosis, pyrotopsis, and necrosis: Mechanistic description of dead and dying eukaryotic cells, *Infection and Immunity*, 73, 1907–1916, 2005.

16. M. Archana, T. Yogesh and K. Kumaraswamy, Various methods available for detection of apoptotic cells-a review, *Indian Journal of Cancer*, 50, 274–283, 2013.

17. Y. Soini, P. Pääkkö and V. Lehto, Histopathological evaluation of apoptosis in cancer, *The American Journal of Pathology*, 153, 1041–1053, 1998.

18. X. M. Sun, M. MacFarlane, J. Zhuang, B. B. Wolf, D. R. Green and G. M. Cohen, Distinct caspase cascades are initiated in receptor-mediated and chemical-induced apoptosis, *Journal of Biological Chemistry*, 274, 5053–5060, 1999.

19. S. E. Logue and S. J. Martin, Caspase activation cascades in apoptosis, *Biochemical Society Transactions*, 36, 1–9, 2008.

20. W. R. Duan, D. S. Garner, S. D. Williams, C. L. Funckes-Shippy, l. S. Spath and E. A. Blomme, Comparison of immunohistochemistry for activated caspase-3 and cleaved cytokeratin 18 with the TUNEL method for quantification of apoptosis in histological sections of PC-3 subcutaneous xenografts, *The Journal of Pathology*, 199, 221–228, 2003.

21. A. Verhoven, S. Krahling, R. A. Schlegel and P. Williamson, Regulation of phosphatidylserine exposure and phagocytosis of apoptotic T lymphocytes, *Cell Death & Differentiation*, 6, 262–270, 1999.

22. I. Vermes, C. Haanen, H. Steffens-Nakken and C. Reutellingsperger, A novel assay for apoptosis flow cytometric detection of phosphatidylserine expression on early apoptotic cells using fluorescein labelled Annexin-V, *Journal of Immunological Methods*, 184, 39–51, 1995.

23. A. Petrovsky, E. Schellenberger, L. Josephson, R. Weissleder and A. Bogdanov, Near-infrared fluorescent imaging of tumor apoptosis, *Cancer Research*, 63, 1936–1942, 2003.

24. F. Oberhammer, J. Wilson, C. Dive, I. Morris, J. Hickman, A. Wakeling, P. R. Walker and M. Sikorska, Apoptotic death in epithelial cells: Cleavage of DNA to 300 and/or 50 kb fragments prior to or in the absence of internucleosomal fragmentation, *The EMBO Journal*, 12, 3679, 1993.

25. N. P. Singh, A simple method for accurate estimation of apoptotic cells, *Experimental Cell Research*, 256, 328–337, 2000.

26. P. Salgame, L. L. Primiano, J. E. Fincke, S. Muller and M. Monestier, An ELISA for detection of apoptosis, *Nucleic Acids Research*, 25, 680–681, 1997.

27. S. Mundle, X. Gao, S. Khan, S. Gregory, H. Preisler and A. Raza, Two *in situ* labeling techniques reveal different patterns of DNA fragmentation during spontaneous apoptosis *in vivo* and induced apoptosis *in vitro*, *Anticancer Research*, 15, 1895–1904, 1994.

28. K. Kyrylkova, S. Kyryachenko, M. Leid and C. Kioussi, Detection of apoptosis by TUNEL assay, *Odontogenesis: Methods and Protocols*, 41–47, 2012.
29. A. R. Collins, V. L. Dobson, M. Dušinská, G. Kennedy and R. Štětina, The comet assay: What can it really tell us? *Mutation Research/Fundamental and Molecular Mechanisms of Mutagenesis*, 375, 183–193, 1997.
30. L. Zamai, E. Falcieri, G. Zauli, A. Cataldi and M. Vitale, Optimal detection of apoptosis by flow cytometry depends on cell morphology, *Cytometry*, 14, 891–897, 1993.
31. K. Liu, P. C. Liu, R. Liu and X. Wu, Dual AO/EB staining to detect apoptosis in osteosarcoma cells compared with flow cytometry, *Medical Science Monitor Basic Research*, 21, 15–20, 2015.
32. A. Ribble, N. B. Goldstein, D. A. Norris and Y. G. Shellman, A simple technique for quantifying apoptosis in 96-well plates, *BMC Biotechnology*, 5, 12–19, 2005.
33. S. W. Perry, J. P. Norman, J. Barbieri, E. B. Brown and H. A. Gelbard, Mitochondrial membrane potential probes and the proton gradient: A practical usage guide, *Biotechniques*, 50, 98–115, 2011.
34. M. C. Pinto, D. F. Dias, H. L. Del Puerto, A. S. Martins, A. Teixeira-Carvalho, O. A. Martins-Filho, B. Badet, P. Durand, R. J. Alves and E. M. Souza-Fagundes, Discovery of cytotoxic and pro-apoptotic compounds against leukemia cells: Tert-butyl-4-[(3-nitrophenoxy) methyl]-2, 2-dimethyloxazolidine-3-carboxylate, *Life Sciences*, 89, 786–794, 2011.
35. S. Sahara, M. Aoto, Y. Eguchi, N. Imamoto, Y. Yoneda and Y. Tsujimoto, Acinus is a caspase-3-activated protein required for apoptotic chromatin condensation, *Nature*, 401, 168–173, 1999.
36. S. A. Yoo, H. J. Yoon, H. S. Kim, C. B. Chae, S. De Falco, C. S. Cho and W. U. Kim, Role of placenta growth factor and its receptor FLT-1 in rheumatoid inflammation: A link between angiogenesis and inflammation, *Arthritis Rheum*, 60, 345–354, 2009.
37. S.-W. Yu, S. A. Andrabi, H. Wang, N. S. Kim, G. G. Poirier, T. M. Dawson and V. L. Dawson, Apoptosis-inducing factor mediates poly (ADP-ribose)(PAR) polymer-induced cell death, *Proceedings of the National Academy of Sciences*, 103, 18314–18319, 2006.
38. I. Nicoletti, G. Migliorati, M. Pagliacci, F. Grignani and C. Riccardi, A rapid and simple method for measuring thymocyte apoptosis by propidium iodide staining and flow cytometry, *Journal of Immunological Methods*, 139, 271–279, 1991.
39. W. G. Telford, L. E. King and P. J. Fraker, Comparative evaluation of several DNA binding dyes in the detection of apoptosis-associated chromatin degradation by flow cytometry, *Cytometry*, 13, 137–143, 1992.
40. A. Ferlini, S. Di Cesare, G. Rainaldi, W. Malorni, P. Samoggia, R. Biselli and A. Fattorossi, Flow cytometric analysis of the early phases of apoptosis by cellular and nuclear techniques, *Cytometry*, 24, 106–115, 199.

6 Cell Migratory Assays

6.1 INTRODUCTION

Metastasis is a cascade of events that results in the spread of a malignant tumor from the primary site of origin into different organs of the body. This malignancy progression involves numerous complex processes such as alterations in the adhesive properties of tumor cells and their interactions with the extracellular matrix (ECM), modifications in the expression of ECM proteins, infiltrative progression through the ECM, the migration of cells through blood or lymphatic vessels, and increased number of distant colonies [1]. These variations are responsible for the spread, survival, and growth of cancer in a secondary site ensuing metastases and have consequently attracted considerable interest in anticancer therapy [2]. As most of the cancer-related deaths are ascribed to metastatic dissemination, biochemical processes both at molecular and cellular levels fundamental to metastasis remain to be a major thrust area of cancer research. The development of novel therapeutic strategies to avoid metastatic spread remains as a challenge for both medicinal chemists and biologists [3].

The understanding of various factors and complex molecular events regulating metastatic events has paved the way to the development of a broad spectrum of two dimensional (2D) and three-dimensional (3D) *in vitro* assays to mimic in part or entirely the several steps in metastatic cascade under a more meticulous environment. These assays are highly valued in cancer research not only as techniques to explain the molecular events that support metastasis, but also to permit drug screening and validating therapeutic targets [4].

Migration refers to any directed movement of cell inside the body which permits the cells to alter their position either within tissues or between other organs. Poorly differentiated malignant cells undergo epithelial to mesenchymal transition (EMT) [5]. The ability of these transformed cells to attach strongly to the ECM, cytoskeletal contractibility, and protracted spindle-like cell bodies [6] aids in the cell migration process. Traditionally in 2D *in vitro* assays, tumor-matrix interactions have been studied by seeding the tumor cells into ECM proteins (e.g., collagen) coated culture plates. The receptors involved are explored using specific adhesion receptor inhibitors or function-blocking antibodies and resistance to cell adhesion is quantified typically either through cell counting or using fluorometric (e.g., calcein AM) or colorimetric (e.g., crystal violet) dyes.

This chapter gives an overview of *in vitro* bioassays that assess the migratory potential of cells and focuses on the commonly used experimental protocols. However, to date, none of the defined assays could reasonably reiterate all crucial steps involved in the cell migratory aspects of metastasis, but only recapitulate parts of it. In all these assays, though several variations can be implemented, the main attention is on the culture conditions selected; notably the ECM protein used must be pertinent to the tumor type being investigated.

6.2   2D CELL MIGRATION ASSAYS

The 3D movement of cells which are embedded in the ECM in our body is fundamental to life and varies in their respective composition, density, and stiffness. These physiological circumstances are represented by 3D cell migration assays. However, 2D migration assays are still required and performed to accomplish comprehensive understanding on the influence of (extra) cellular proteins and matrix-components on cell movements. Besides, there are several new perspectives on the use of modified protocols and development of novel 2D cell migration assays [7].

Collective cell migration plays an integral role in metastasis of malignant tumors. Malignant cells migrate either directionally (taxis) when the external signal is conveyed in the form of a gradient or move randomly (kinesis) when a homogeneous concentration of soluble growth factors (chemokinesis) exists or in the presence of an adhesive substrate (haptokinesis). During the process of collective migration, the cells frequently form unified monolayers and interact both biochemically and mechanically via intercell adhesions. Cell signaling, proliferation, as well as cell to microenvironment interaction also play a vital role in this complex process. Various 2D and 2D/3D *in vitro* procedures have been established to study the dynamic collective cell migration process.

6.2.1   WOUND HEALING ASSAY

The ability to visualize the cells during their migration ranks *in vitro* wound-healing assay as one of the most common and earliest methods in measuring collective cell migration. The assay has a simple protocol and can be applied in the majority of the cell types available [8,9]. Typical wound healing assay measures the 2D cell migration into a wound (cell-free area) which is generated by a central thin linear scratch across the surface of a dense confluent monolayer of cells confined in a tissue culture plate (Figure 6.1). The wound region created stimulates various cellular responses,

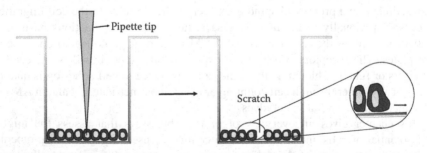

FIGURE 6.1   Schematic diagram of wound healing assay; scratch-off cells from a dense monolayer on ECM-coated or glass or plastic surfaces to produce a cell-free area using a vertically held pipette tip to observe cell migration followed by wound closure.

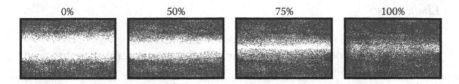

| 0% | 50% | 75% | 100% |

FIGURE 6.2 Percentage of wound closure observed under a microscope.

mainly collective cell migration. The dynamics of cell migration toward this gap (healing) or the cell growth toward the center of the gap is quantitated and examined using time-lapse microscopy (Figure 6.2). The kinetics of cell migration for adherent cells in horizontal direction can also be detected using digital and fluorescence imaging at various time points in real time.

Procedure:

1. Seed required number of cells into 6-, 12-, or 24-well tissue culture plate in DMEM supplemented with 10% FBS to obtain ~70%–80% confluence monolayer after 24 h.
2. Gently scratch a straight line in one direction in the monolayer across the center of the plate with a sterile 1 mL pipette tip held perpendicular to the bottom of the plate.
3. Wash the plate twice gently with the medium to take away the detached cells.
4. Replenish the plate with fresh medium.
5. Treat the cells with the test compound.
6. Grow cells for additional 24 h.
7. Wash the cells twice with PBS.
8. Fix the cells using 3.7% paraformaldehye for about 30 min.
9. Stain using 1% crystal violet solution prepared in 2% ethanol.
10. Incubate for another 30 min.
11. Image the stained monolayer with a microscope or measure the change in gap width.

Advantages:

- Simple, popular, and technically nondemanding technique.
- Fast, convenient, and inexpensive.
- No special equipment needed.
- Any ECM-coated plate can be used.
- Rapid setup, easy readout and analysis.
- Easily adjustable experimental conditions for different purposes.
- Real-time quantification of cell migration kinetics achievable.
- High-throughput screening process feasible.
- Automated data collection available.
- Fluorescent labeling of cell components possible.
- Able to evaluate cell to ECM interactions.

Limitations:

- Scratch may be uneven, not of uniform width.
- Variations in wound width preceding cell migration are critical as the speed of cell migration just before wound closure normally increases.
- Scratching can often damage cells and the ECM coating beneath.
- Inconsistent scratch size, injured cells, and damaged ECM affect reproducibility.
- Incorrect results arise when some cells keep adhered to the margin of the scratch after wounding, which can sometimes later reattach to the plate and migrate toward the wounded area.
- Lack of standardization between labs.
- Not suitable for nonadherent cells and chemotaxis.
- Wound healing assay performed for a higher duration (>24 h) cannot distinguish cell proliferation and changes in cell survival from cell motility.

6.2.2 CELL EXCLUSION ZONE ASSAY

In cell exclusion or chemokinesis assays, cells are seeded into wells and a cell-free region is generated by an insert or a barrier, which is removed once a confluent monolayer surrounds it. The 2D void closure is monitored when the cells migrate into the cell-free zone and can be measured by photomicrography (Figure 6.3). Also, the cells can be covered using a mask prior to the experiment and the migrated cells, if fluorescently labeled, can be quantified as they leave the masked area using microplate readers. The fluorescence signal detected will be proportional to the extent of migrated cells. Cell exclusion assays are beneficial over scratch assays as the presence of residual cell debris is reduced and reproducibility is better [10–12].

Procedure:

1. Position the stoppers in the well to generate a cell exclusion zone.
2. Seed 5×10^4 cells in each well and incubate for 24 h.
3. Remove the stopper to create a 2 mm diameter circular cell-free area.

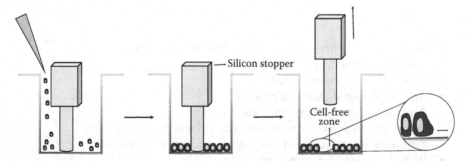

FIGURE 6.3 Schematic diagram of cell exclusion zone assay; a cell-free zone is created using silicone stoppers, which are removed during the commencement of the testing process and migration is studied.

4. Wash the cells carefully with medium to remove any floating cells without disturbing the adhered cells.
5. Treat the cells with the drug; incubate for 24 h to allow cell migration into the detection zone.
6. Image the cells during premigration and after the cells migrate inward to fill the gap.

Advantages:

- Cell-free zones are precisely defined voids with reproducibly similar sizes and sharp margins.
- No damaged cells due to mechanical scraping or electric ablation.
- Well-standardized convenient procedure which is fast and inexpensive.
- Simple technical setup and does not need special equipment for investigation.
- Real-time quantification of cell migration kinetics possible.
- Amenable to high-throughput screening.
- Possible to label cell constituents with fluorescent dyes.
- Automated data acquisition obtainable.
- Able to assess interactions between tumor cell and ECM.

Limitations:

- Only suited for adherent cells.
- Not suitable for chemotaxis.
- Cells may enter the cell exclusion region beneath the stoppers if they are not firmly attached.
- Incorrect results are obtained if floating cells attach in the cell exclusion area.

6.2.3 TRANSWELL MIGRATION ASSAY

Cell migration assays performed in Transwell chambers are also referred to as Boyden Chamber Assays. This is an endpoint assay that measures the single cell migration movement first in the horizontal and then in the vertical directions. The cells are seeded into the top chamber and the migration of cells to a bottom chamber, which is separated by a microporous membrane is monitored (Figure 6.4). The cells move in response to a chemotactic gradient produced by the addition of either serum or explicit chemotactic factors into the lower compartment [13–15].

Procedure:

1. Transwell plates with permeable supports with/without ECM coating are to be used for the study. The plates contain two compartments, upper and lower in a well.
2. Add medium with serum or any specific chemotactic factors to the lower compartment.
3. Seed the cells in the upper chamber and incubate for a day.

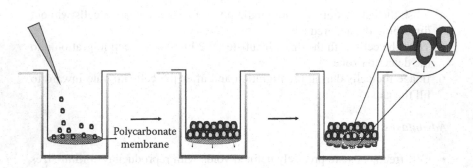

FIGURE 6.4 Schematic diagram of Transwell Migration Assay; depicts cell migration through a porous membrane.

4. Treat the cells with test compound keeping appropriate controls.
5. Allow the cells to migrate through the membrane to the lower well. The proper incubation time varies with different cell types and is determined until the motile cells appear at the other side of the filter.
6. Fix the membrane and remove the nonmigrated cells from the upper side of the filter using a cotton swab.
7. Stain the migrated cells with cytological dyes like crystal violet or fluorescent dyes.
8. Count the number of migrated cells on the undersurface of the separating membrane using a fluorescent reader.

Advantages:

- The migratory behavior of both adherent and nonadherent cells can be studied.
- Fast and often completed in 3–4 h.
- Appropriate assay to measure haptotaxis or chemotaxis.
- Able to evaluate tumor cell and ECM interactions.
- Migrating cells can be recovered.
- No special equipment needed.
- Technically nondemanding.

Limitations:

- Difficult to control the steepness of the chemotactic gradient and maintain it over longer duration.
- Counting the number of migrated cells consumes more time (to avoid this, precoated Transwell plates can be used).
- Kinetic analysis of cell migration is not possible as it has a set endpoint.

6.2.4 Fence Assay

The principle of fence or ring assay is the reverse of that of the cell exclusion zone assay. The fence assay differs from the wound healing assay; wherein it prevents the

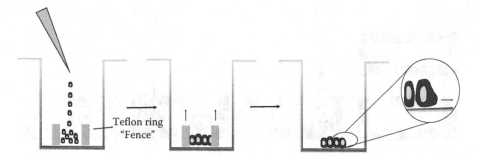

FIGURE 6.5 Schematic diagram of fence assay; cells are seeded inside a ring-shaped plastic device and cell movement after the removal of the ring to the cell-free surrounding zone is measured.

cells from actually growing on the ring/fence while a confluent layer develops inside the ring (Figure 6.5). The migration of cells outward to the cell devoid area on the removal of the cylindrical plastic fence is measured [16–18].

Procedure:

1. Place either Teflon, metal, or glass ring (fence) on a standard cell culture dish.
2. Seed the cells into the inner area of the fence encircled by the ring device.
3. Restrict the cell attachment zone within the void inside the ring.
4. Detach the ring and wash gently to remove the nonattached cells.
5. Incubate the cells with test compounds.
6. Allow cells to migrate radially outward from the circular area.
7. Measure the cell movement using a digital automated image analyzer as the increase in the circumference due to outward migration of cells.

Advantage:

• Technically nondemanding technique.

Limitations:

• Applicable for adherent cells only.
• Custom-made rings essential.

6.2.5 MICROFLUIDIC ASSAY

Microfluidic or capillary chamber migration assays involve seeding cells into a horizontal setting of two chambers that are connected sidewise by a narrow internal channel. The entire device is roofed by a glass slide, and the system provides two ports for medium delivery. One of the compartments is filled with cells suspended in the medium, while the other is loaded with medium comprising a chemoattractant. A stable concentration gradient develops in the bridging capillary area located

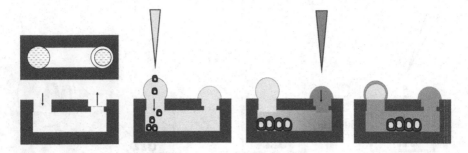

FIGURE 6.6 Schematic diagram of microfluidic assay; cells travel along a thin bridging capillary in response to a steady gradient of a chemoattractant.

between the two reservoir chambers, and the kinetics of horizontal cell migration is studied (Figure 6.6). The number of migrating cells on the surface of the capillary is counted by light microscopy [19–21].

Procedure:

1. Introduce either cells alone or suspended in a matrix into the smaller port.
2. Allow cells to adhere to the bottom of the chamber.
3. Expose the cells to the test agent.
4. Add the medium containing the chemoattractant to the larger port.
5. Allow a gradient to establish based on surface tension.
6. Image the cells or measure cell migration.

Advantages:

- Easy handling and technically nondemanding.
- Assesses directed chemotaxis along a gradient.
- Allows analysis of multiple chemoattractants.
- Measures chemotaxis of known linear chemotactic gradients.
- Appropriate for rare cell types and to test expensive compounds as only small volumes are needed for the assay.
- Ideal for live cell imaging.
- Offers data on the directionality and speed of cell migration.
- Constant gradient for up to 48 h.
- Can be used for both adherent and cell suspension.

Limitations:

- Usage of less volume of the media demands more frequent media changes and cautious humidification of the incubator.
- Expensive.
- Difficult to set up automation processes with the existing systems.
- Most systems do not meet high-throughput screen requirements.
- Quantification using imaging techniques is challenging.

6.2.6 SINGLE CELL MOTILITY ASSAY

Colloidal gold particle coated surfaces can be employed to study single cell migration by following their individual cell paths [22]. Cells are seeded at low density onto gold particle-coated tissue culture plates. The colloidal gold particles are visualized as a uniform layer of small dark dots under the microscope. The assay is based on the ability of the migrating single cells to phagocytize the gold particles and remove them from the culture plates. This results in white tracks that can be photographed or the cleared areas can be quantitatively evaluated (Figure 6.7). The assay can also be performed by replacing gold particles by quantum dots.

Procedure:

1. Place the sterile coverslips coated with colloidal gold particles in an appropriate cellular media.
2. Treat around 1×10^3 cells/mL with the test drug.
3. Transfer the treated cells onto the gold-coated coverslips kept in a 24-well dish.
4. Incubate at 37°C for desired period depending on the cells studied.
5. Capture the images of the trails produced by a single migrating cell using a light microscope.

Advantages:

• Suitable for automated and high-throughput screening assays.
• Possible to track single cell movement.
• Can monitor chemokinesis or undirected movement.
• Possible to detect real-time path.
• Can determine absolute migration speed.
• Simple, technically nondemanding, and simple readout.

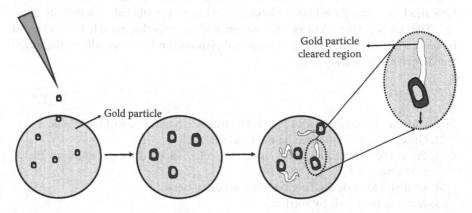

FIGURE 6.7 Schematic diagram of single cell motility assay; cells remove the gold particles on their migration path leaving behind bright trails.

Limitations:

- Small sample size based on one cell–one track principle.
- Extreme manual labor or else completely automated system required.

Though there are various alternatives to measure *in vitro* cell migration, the above surrogate 2D migration assays depict a simplified version of the actual metastatic cascade taking place *in vivo*. Hence, the significance of data collected via these tests must be confirmed *in vivo*.

6.3 2D/3D MIGRATION ASSAYS

The 2D monolayer cell cultures afford a convenient and swift platform to gain knowledge on cell adhesion and migration. However, these models lack the intricate architectural complexity of malignant tumors *in vivo*. 3D multicellular culture systems provide a more suitable physiological microenvironment than 2D monolayer cultures to study tumor progression *in vitro* and for predicting drug sensitivity *in vivo* as well as the identification of appropriate therapeutic targets. In contrast to nontumorigenic epithelial cells, distinguishing features of tumor cells like the propensity to undergo self-aggregation in the absence of an adhesive substrate add to their metastatic potential *in vivo* [23,24]. Approaches that take advantage of the characteristic feature of cancer cells to produce multicellular aggregates/spheroids have been employed in 3D migration assays. These assays are considered to be more physiologically relevant and better simulate the complex topography and architecture of solid tumors *in vivo*.

6.3.1 Microcarrier Bead Assay

Measurement of cell mobility from microcarrier beads onto the surfaces of 2D cell culture vessel forms the basis of this assay [25]. Microcarrier bead faces are coated with cells which further grow to form a confluent layer on the bead surface. Later these beads are transferred into cell culture dishes and incubated for a definite duration. The beads are then detached by suction, and the cells that move to the surface of the cell culture vessel are fixed, stained, and estimated microscopically or measured densitometrically (Figure 6.8).

Procedure:

1. Coat microcarrier beads (DEAE Dextran or Cytodex beads) with cells.
2. Grow cells to confluence on the bead surface.
3. Place the cell-coated beads onto the cell culture dishes containing the medium.
4. Incubate for a desired period with test compounds.
5. Remove the beads by suction.
6. Fix, stain, and evaluate the surface area of cell migration microscopically or densitometrically.

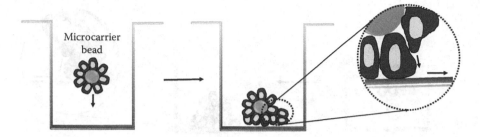

FIGURE 6.8 Schematic diagram of microcarrier bead assay; cells from microcarrier beads migrate radially and the area can be measured.

Advantages:

• Cells establish intimate cell to cell interactions on the bead surface, more closely imitating the tight contact present in cells *in vivo*.
• Restricted space and little cell size difference ensure a fairly constant number of cells on the bead surface when confluence is attained.
• Can monitor cell coating effortlessly using a conventional light microscope.

Limitations:

• The microbeads are expensive.
• Can be employed for adherent cells only.
• Technically medium demanding.
• Beads that are not or inadequately covered by cells need to be identified and removed from analysis using a conventional light microscope.

6.3.2 SPHEROID MIGRATION ASSAY

This assay principle has close similarity to that of microcarrier bead assay and is a combination of 3D and 2D technologies. Multicellular tumor cell spheroids formed from a certain cell type under study are placed on top of a cell culture dish. The cells start to migrate soon after adhesion of the spheroid onto the culture dish surface (Figure 6.9). The zone of attachment increases concentrically as the cells migrate out and finally the cell migration can be measured under a microscope [26,27].

6.3.2.1 3D-Spheroid Formation Protocol

3D models enable more relevant and translational observations compared to 2D models. Briefly, 1–5 K cells in suspension are added per well to an ultra-low attachment (ULA) plate and within 48–96 h a single spheroid forms in each well. This generic method can be applied to >25 different cancer cell types.

1. Collect freshly confluent (~80%) T75 flask of cells, remove media, and wash cells with 10 mL of D-PBS.

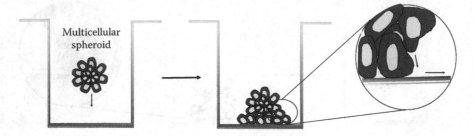

FIGURE 6.9 Schematic diagram of spheroid migration assay; cells from multicellular spheroids move outward concentrically and the rise in dissemination area can be studied over time.

2. Detach the cells by adding 1 mL Trypsin/EDTA, leave for 4–5 min.
3. Add 9 mL culture media, and resuspend the cells.
4. Count the cells (e.g., Trypan blue staining + hemocytometer), centrifuge the cell suspension (1000 rpm, 4 min) and resuspend the cell pellet in culture media at 2.5×10^4 cells/mL.
5. Seed 2500 cells per well into each well of a ULA plate, centrifuge the plate (1000 rpm, 10 min) at room temperature, and allow the cell plate to equilibrate.
6. Monitor spheroid formation over 72 h until the desired spheroid size is achieved.

Procedure:

1. Wash tumor cell type monolayers with PBS, add cell dissociation enzyme (1 mL for T25 and 2 mL for T75 flask), and incubate cells at 37°C for 2–5 min.
2. Check for cell detachment under a microscope and neutralize cell dissociation enzyme with DMEM (5 mL for T25 or 8 mL for T75 flask).
3. Centrifuge cell suspension at $500 \times g$ for 5 min, remove supernatant, and resuspend cell pellet in 1 mL of DMEM to get a single cell suspension without cell clusters.
4. Count cells and dilute the cell suspension to obtain $0.5-2 \times 10^4$ cells/mL.
5. Transfer 200 µL cell suspension to each well in a well plate, incubate at 37°C, 5% CO_2, 95% humidity for 4–5 days for tumor spheroid formation.
6. Transfer the spheroid onto conventional tissue culture dishes. The spheroids attach to the surface and cells start to move concentrically outward.
7. Measure cell migration or the spreading area microscopically.

Advantages:

• Can achieve better physiologic tissue-like morphology with close cell to cell association by the usage of 3D spheroid structures.

- Maintain diverse cellular statuses as oxygen and nutrient supply is concerned as in *in vivo* condition.
- Mimics cell migration out of small malignant clusters closely.

Limitations:

- Applicable only to the cells that are capable of forming spheroids.
- Suspension culture cannot be studied.
- Technically medium-demanding.

REFERENCES

1. G. Gupta and J. Massagué, Cancer metastasis: Building a framework, *Cell*, 127(4), 679–695, 2006.
2. J. Chia, N. Kusuma, R. Anderson, B. Parker, B. Bidwell, L. Zamurs, E. Nice and N. Pouliot, Evidence for a role of tumor-derived Laminin-511 in the metastatic progression of breast cancer, *The American Journal of Pathology*, 170(6), 2135–2148, 2007.
3. G. Bendas and L. Borsig, Cancer cell adhesion and metastasis: Selections, Integrins, and the Inhibitory Potential of Heparins, *International Journal of Cell Biology*, 1–10, 2012.
4. N. Kramer, A. Walzl, C. Unger, M. Rosner, G. Krupitza, M. Hengstschläger and H. Dolznig, *In vitro* cell migration and invasion assays, *Mutation Research/Reviews in Mutation Research*, 752(1), 10–24, 2013.
5. J. Thiery, Epithelial–mesenchymal transitions in tumour progression, *Nature Reviews Cancer*, 2(6), 442–454, 2002.
6. F. Grinnell, Fibroblast biology in three-dimensional collagen matrices, *Trends in Cell Biology*, 13(5), 264–269, 2003.
7. R. V. Horssen and T. T. L. M. Hagen, Crossing barriers: The new dimension of 2D cell migration assays, *Journal of Cellular Physiology*, 226(1), 288–290, 2010.
8. C. Liang, A. Park and J. Guan, *In vitro* scratch assay: A convenient and inexpensive method for analysis of cell migration *in vitro*, *Nature Protocols*, 2(2), 329–333, 2007.
9. L. G. Rodriguez, X. Wu and J-L. Guan, Wound-healing assay, *Cell Migration: Developmental Methods and Protocols*, 294, 23–29, 2005.
10. M. Poujade, E. Grasland-Mongrain, A. Hertzog, J. Jouanneau, P. Chavrier, B. Ladoux, A. Buguin and P. Silberzan, Collective migration of an epithelial monolayer in response to a model wound, *Proceedings of the National Academy of Sciences*, 104(41), 15988–15993, 2007.
11. T. Omelchenko and A. Hall, Myosin-IXA regulates collective epithelial cell migration by targeting RhoGAP activity to cell-cell junctions, *Current Biology*, 22(4), 278–288, 2012.
12. A. Fougerat, N. Smirnova, S. Gayral, N. Malet, E. Hirsch, M. Wymann, B. Perret, L. Martinez, M. Douillon and M. Laffargue, Key role of PI3Kγ in monocyte chemotactic protein-1-mediated amplification of PDGF-induced aortic smooth muscle cell migration, *British Journal of Pharmacology*, 166(5), 1643–1653, 2012.
13. W. Chen, K. Kuo, T. Chou, C. Chen, C. Wang, Y. Wei and L. Wang, The role of cytochrome c oxidase subunit Va in non-small cell lung carcinoma cells: Association with migration, invasion and prediction of distant metastasis, *BMC Cancer*, 12(1), 273, 2012.
14. R. Harisi, I. Kenessey, J. N. Olah, F. Timar, I. Babo, G. Pogany, S. Paku and A. Jeney, Differential inhibition of single and cluster type tumor cell migration, *Anticancer Research*, 29(8), 2981–2985, 2009.

15. S. Qi, Y. Song, Y. Peng, H. Wang, H. Long, X. Yu, Z. Li et al., ZEB2 mediates multiple pathways regulating cell proliferation, migration, invasion, and apoptosis in glioma, *PLoS ONE*, 7(6), e38842, 2012.
16. G. Cai, J. Lian, S. S. Shapiro and D. A. Beacham, Evaluation of endothelial cell migration with a novel *in vitro* assay system, *Methods in Cell Science*, 22(2–3), 107–114, 2000.
17. S. Sagnella, F. Kligman, E. Anderson, J. King, G. Murugesan, R. Marchant and K. Kottke-Marchant, Human microvascular endothelial cell growth and migration on biomimetic surfactant polymers, *Biomaterials*, 25(7–8), 1249–1259, 2004.
18. W. Elbjeirami and J. West, Angiogenesis-like activity of endothelial cells co-cultured with VEGF-producing smooth muscle cells, *Tissue Engineering*, 12(2), 381–390, 2006.
19. S. H. Zigmond, Orientation chamber in chemotaxis, *Methods in Enzymology*, 162, 65–72, 1987.
20. D. Zicha, G. A. Dunn and A. F. Brown, A new direct-viewing chemotaxis chamber, *Journal of Cell Science*, 99(4), 769–775, 1991.
21. S. Chaubey, A. J. Ridley and C. M. Wells, Using the Dunn chemotaxis chamber to analyze primary cell migration in real time, *Cell Migration: Developmental Methods and Protocols*, 769, 41–51, 2011.
22. Y. Niinaka, A. Haga and A. Raz, Quantification of cell motility, *Metastasis Research Protocols: Volume II: Analysis of Cell Behavior In Vitro and In Vivo*, 58, 55–60, 2001.
23. J. Rak, Massive programmed cell death in intestinal epithelial cells induced by three-dimensional growth conditions: Suppression by mutant c-H-ras oncogene expression, *The Journal of Cell Biology*, 131(6), 1587–1598, 1995.
24. S. Kim, J. Turnbull and S. Guimond, Extracellular matrix and cell signalling: The dynamic cooperation of integrin, proteoglycan and growth factor receptor, *Journal of Endocrinology*, 209(2), 139–151, 2011.
25. E. Rosen, L. Meromsky, E. Setter, D. Vinter and I. Goldberg, Quantitation of cytokine-stimulated migration of endothelium and epithelium by a new assay using microcarrier beads, *Experimental Cell Research*, 186(1), 22–31, 1990.
26. S. Konduri, A. Tasiou, N. Chandrasekar and J. Rao, Overexpression of tissue factor pathway inhibitor-2 (TFPI-2), decreases the invasiveness of prostate cancer cells *in vitro*, *International Journal of Oncology*, 18, 127–131, 2001.
27. M. M. Knupfer, F. Pulzer, I. Schindler, P. Hernaiz Driever, H. Knupfer and E. Keller, Different effects of valproic acid on proliferation and migration of malignant glioma cells *in vitro*, *Anticancer Research*, 21, 347–351, 2000.

7 *In Vitro* Cell Invasion Assays

7.1 INTRODUCTION

Tumor cells attain the capability to invade into the adjacent surrounding tissue and spread into far off organs leading to metastasis as malignancy advances. During the invasion process, the cancer cells form protrusive structures or invadopodia which aid them to penetrate into the basement membrane and mediate cell attachment and remodeling of the ECM [1,2]. The invadopodia and podosomes are rich in actin fibers, proteinases, and specific adhesion proteins and are jointly accountable for tumor cell mobility as well as ECM degradation [3]. Invadopodia spread into the ECM and enable invasion, extravasation into the vascular canals, allowing either hematogenous or lymphatic propagation and subsequently metastasis.

The need to identify appropriate inhibitors to cancer invasion has facilitated the development of various quantitative *in vitro* invasion assays [4]. Though cell invasion partly resembles cell migration, it mainly requires a cell to move through an ECM or basement membrane extract (BME) barrier by enzymatically degrading it initially and then later to get established on a fresh site. An extensive range of natural BMEs such as human amniotic membranes132 or chick chorioallantoic (CAM) membranes133 has been developed and used in these *in vitro* invasion assays to mimic the *in vivo* situations. The *in vitro* invasion approaches that are developed offer a fast and more economical alternative to the use of animal models and are amenable to interventions that are not practical *in vivo*. However, these *in vitro* models lack reproducibility due to the intrinsic heterogenic nature of tissue preparations.

In *in vitro* techniques, invasion is often quantified based on the number of invading cells and/or the distance moved by the cells from the surface of the gel [5]. Malignant cells can be entirely implanted into the matrix either as a single cell suspension or as spheroids. They are positioned between the two stratum of the ECM gel, and the cells are allowed to move out of the tumor mass and invade into the neighboring matrix [6].

7.2 TRANSWELL INVASION ASSAY

The most frequently employed method to measure tumor cell invasion is Transwell assay which uses well-defined 3D matrices like collagen or matrigel. The main charm of this assay is related to the invasive behavior of tumor cells that correlates well with the actual metastatic scenario in animal models [7,8]. Transwell invasion assays are superior to the standard 2D migration assays because the single cell suspensions are seeded on top or rooted within a filter layered with a dense film of ECM-derived 3D

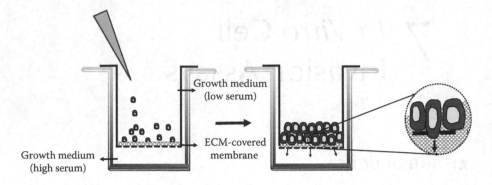

FIGURE 7.1   Schematic diagram of Transwell invasion assay; invasion of the cells through the ECM gel.

protein matrix, typically collagen-I or matrigel (BMM, a basement membrane-like matrix) that imitates the regular composition of basement membranes. The ECM blocks the membrane pores, obstructing the migration of noninvasive cells, whereas the invasive cells are capable of degrading the matrix and traveling through the ECM coating and finally sticking to the bottom of the filter in response to a chemo-attractant. The cells that have invaded the ECM gel and drifted to the underside of the porous membrane are counted after ~18–24 h incubation (Figure 7.1). The invading cells are visualized in the 3D matrix, and the cells need not cross an arti-ficial porous membrane. The extent of invasion is determined quantitatively under a microscope by measuring the distance of migration from the center line in response to the specific ECM proteins embedded in the upper and lower gels [9,10]. The rela-tive involvement of invasion to the speed of cell motility or the invasive index is calculated as the ratio of invaded cells that passed through the ECM-coated filters to the migrated cells (noncoated filters).

Procedure:

1. Transwell polycarbonate inserts that are commercially available can be used for the study. Inserts with different diameter and pore size membranes with/without protein coating are available.
2. Place the inserts in a suitable microtiter plate.
3. Seed the cells and the medium into the upper chamber of the device.
4. Treat the cells with the test agent for 24 h.
5. Detect the invasive cells by staining, light microscope counting, or fluorimetry.

Advantages:

- Can be used for both suspension and adherent cells.
- Does not require any special equipment.
- Relatively simple and easy experimental setup.

- Technically nondemanding.
- Rapid, quantitative, and reproducible results.
- Commercial availability of inserts.
- Provides information on the morphological changes associated with invasive response.
- Quantification by fluorescent dyes gives reliable results.
- Short to medium chemokine gradient maintained between the top cell culture insert and the bottom culture vessel growth medium.

Limitations:

- Difficult to remove noninvaded cells present on the top side of the Transwell insert, before proceeding to staining of invasive cells existing at the bottom of the membrane.
- Endpoint assay.

7.3 PLATYPUS INVASION ASSAY

Though the Platypus invasion assay resembles the cell exclusion zone migration assay described in Section 6.2.2 in using small silicone stoppers fitting into 96-well plates, the method setup is relatively different [11]. Cells are seeded on top of a thin surface coated with ECM and are further blanketed by a second denser stratum of ECM. A silicone insert generates a cell-free elimination area in the middle and the invasive cells that are embedded in ECM move from an exterior loop toward the center (Figure 7.2).

Procedure:

1. Coat the bottom surface of individual wells by a thin layer of BME as ECM substrate (BME can be substituted with collagen I as the ECM component).
2. Position the stoppers to construct an exclusion zone.
3. Seed the cells and allow them to adhere to the surface of the first coating of BME.

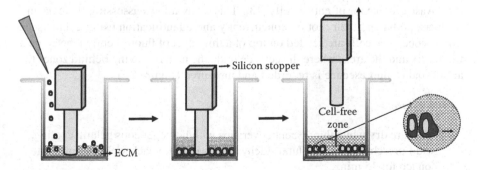

FIGURE 7.2 Schematic diagram of Platypus invasion assay; cells embedded in ECM move from an outer ring toward the cell-free middle zone on the removal of inserts.

4. Treat with test compounds and remove the stopper.
5. Overlay both the cells and the cell-free circular center zone by a denser succeeding layer of BME.
6. Generate a cell layer implanted between the two sheets of ECM and a middle cell-free region filled with ECM.
7. Quantify invasive cell migration into the middle over time with a microscope or fix the cells and analyze the invading cells after immune-fluorescent staining using a confocal microscope.

Advantages:

- Well-standardized and relatively easy assay setup.
- Does not need any specialized equipment.
- Technically nondemanding.
- Possible to execute kinetic analysis.
- Feasible for live imaging of cells.
- Invasion that occurs in the horizontal direction in a thin ECM layer assists assessment using a standard microscope.
- Can study morphology of the invading cell and the gels can be further processed for immunofluorescence analysis.

Limitations:

- No chemokine/growth factor gradient can be formed.
- Real cell to cell interaction is not established prior to the test as the cells are embedded as a single cell suspension.

7.4   GELATIN DEGRADATION ASSAY

Invadopodia are the structures that foster the cancer cells with the ability to invade and metastasize. The invasive capability of the cell can be studied by detecting the formation of invadopodia that degrades the ECM [12]. Gelatin degradation protocol is designed to visualize these cellular protrusions which damage the ECM at a higher resolution and also to quantify invasion at the subcellular level rather than examining the invasive behavior of entire cells [13]. This assay allows assessing the existence of invadopodia and their motion concurrently and quantification using a fluorescent microscope. The cells are seeded on top of a thin layer of fluorescently labeled gelatin matrix and the areas where the cells degrade the matrix leaving behind zones that are devoid of fluorescence is recorded and measured (Figure 7.3).

Procedure:

1. Prepare dry white fluorescent coverslips with homogeneous gelatin coating.
2. Add pre-chilled 1 mL glutaraldehyde solution into each well and incubate on ice for 15 min.
3. Remove glutaraldehyde solution and wash the coverslips at room temperature three times with PBS.

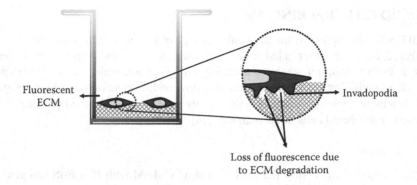

Fluorescent
ECM

Invadopodia

Loss of fluorescence due
to ECM degradation

FIGURE 7.3 Schematic diagram of gelatin degradation assay; loss of fluorescence at locations where invadopodia-mediated fluorescently labeled gelatin matrix degradation occurs.

4. Add freshly prepared 1 mL sodium borohydride solution (5 mg/mL in PBS), keep at room temperature for 3 min, and then stir the plate to avoid any hydrogen bubbles formed.
5. Remove sodium borohydride solution and wash the coverslips three times with PBS at room temperature.
6. Transfer the coverslips to each well of a sterilized 12-well plate and wash three times with PBS.
7. Add 1 mL of cell suspension containing 20–40,000 cells per well containing 1 mL of medium.
8. Treat with the test compound and incubate for 8–16 h.
9. Wash the cells once with PBS at the endpoint and fix rapidly in 4% formaldehyde solution in PBS at room temperature for 10–15 min in the dark.
10. Remove formaldehyde and wash three times with PBS.
11. Incubate in 3% BSA solution in PBS containing 0.1% Triton X-100 at room temperature for 15–30 min in the dark.
12. Remove BSA solution, stain F-actin with phalloidin and incubate for 30–60 min at room temperature in the dark.
13. Remove phalloidin solution, wash three times with PBS.
14. Detect and quantify invadopodia using fluorescence imaging.

Advantages:

- Subcellular invasion areas are detectable.
- Commercially available.

Limitations:

- Cannot follow the cells as a whole during their movement.
- Difficult to detect long-term cellular movement.
- Invadopodia structures may not exactly reflect the 3D reality as cells are adhered to a thin ECM layer and hence can adjust to a 2D cell shape.
- Requires fluorescence microscope.

7.5 3D CELL TRACKING ASSAY

In 3D cell tracking assay, the automated image of the path of moving cells can be analyzed using computer-aided time-lapse video microscopy (Figure 7.4). Though it is difficult to monitor single cells in 3D settings over a period, several strategies including multiphoton or confocal microscopy, widefield fluorescence, and digital holography or contrast enhancing microscopy have been established to trace either labeled or unlabeled cells in 3D matrices [14].

Procedure:

1. Trypsinize and collect the cells by adding DMEM with 10% FBS and gently spin down the cells at 1000 g for 5 min.
2. Resuspend the cell pellet gently in DMEM with 10% FBS.
3. Pipette up and down gently to break down the cell pellet cluster into individual cells.
4. Count the cells and dilute to 50,000 cells/mL in DMEM with 10% FBS.
5. Coat 6-well plates with collagen and incubate overnight at 4°C.
6. Wash plates twice with PBS.
7. Dispense 2 mL of cell suspension into each well and incubate the plate at 37°C in a tissue culture incubator overnight after adding the drug.
8. Start computer-aided, time-lapse video microscopy for automated image analysis to track the path of moving cells.

Advantages:

- Real-time detection of individual trails of cells drifting through 3D structures.
- Nondirected cell movement.
- Possible to determine exactly the deviations in direction and the definite length of the invading cell pathways.

Limitations:

- Can analyze only a limited set of cells at the same time.

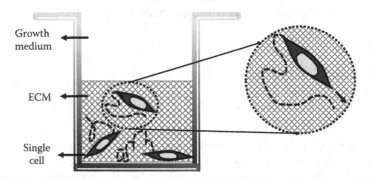

FIGURE 7.4 Schematic representation of 3D cell tracking assay; cells are traced on their course through the ECM.

- Requires special equipment and specialized microscope for live imaging.
- Demands advanced knowledge in data processing.
- Technically highly demanding.

7.6 VERTICAL GEL 3D INVASION ASSAY

This assay is a variant of the vertical invasion assay mainly used for epithelial cell and leukocyte invasion and quantified using radioactive labeled cells or microscopy. The ascending motion of cells from a monolayer, on top of which an ECM layer has been dispensed, is studied. Skin carcinoma cells can be plated on top of a collagen gel layer to monitor the upward invasion from epithelial cell layers (Figure 7.5). Invasion can be measured by immunohistochemical staining, and the depth of invasive areas can be quantified by image analysis software [15].

Procedure:

1. Prepare thick collagen plugs.
2. Plate squamous epithelial cells at the liquid–air interface on top of a layer of collagen gel.
3. Treat the cells with drug and incubate at 37°C, 5% CO_2 for the desired duration.
4. Cells invade vertically into the collagen matrix, which can harbor stromal fibroblasts.
5. Fix cells by adding 400 µL 3% paraformaldehyde in PBS and incubate for 40 min at room temperature and wash with PBS.
6. Permeabilize in 400 µL 0.5% Triton X-100 in PBS for 40 min at room temperature.
7. Wash twice with 400 µL PBS and incubate the cells with 400 µL 1% BSA in PBS for 40 min at room temperature.
8. Quantify invasion by immunohistochemical staining using image analysis software or with formalin-fixed, paraffin-embedded (FFPE) samples, cut perpendicular to the surface.

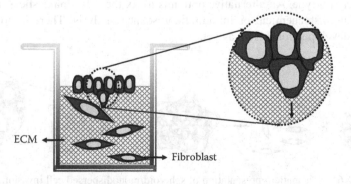

FIGURE 7.5 Schematic representation of vertical gel 3D invasion assay; squamous epithelial cells cultured at the liquid–air boundary invade vertically down into the collagen gel.

Advantages:

- The organotypic skin carcinoma model mostly reproduces the *in vivo* invasion situation.
- It combines 3D invasion into ECM using the heterotypic cell to cell interaction of stromal fibroblasts and epithelial malignant tumor cells.
- Provides information on histological properties as well as cellular morphology.

Limitations:

- Labour intense sample preparation.
- Demands special equipment, such as embedding locations and microtomes as well as formalin-fixed, paraffin-embedded (FFPE) sample cutting experience.
- Requires either both fixation and sectioning or confocal microscope for invasion quantification.
- Endpoint assay.

7.7 SPHEROID/MONODISPERSED CELL INVASION ASSAY

Small clusters of cells formed *in vitro*, called multicellular spheroids, can often mimic *in vivo* 3D tumor aggregates [16,17]. The spheroid/single cell invasion model enables studying the invasive behavior of mostly a nonmalignant cell type A, into a tissue-like assembly comprised of a dissimilar cell form B. In addition, the malignant cell invasion into the spheroid structures composed of nonmalignant or normal cells can also be monitored using this assay. The study is based on the principle that a single cell suspension comprising of cell category A adheres to the co-cultivated spheroid of cell type B and ultimately commences to invade into the spheroid (Figure 7.6). The 3D migration or invasion of fluorescently labeled cells can be analyzed by confocal fluorescence microscopy and the cells invading into the spheroids can be quantified using flow cytometry after the trypsinization process. The study reveals the invasive capability of one cell type to move into tissue-like structures comprised of another cell type. An alternative option is to fix the cells, make slices, and carry out immunohistochemical or immunofluorescence analysis. There is no standard protocol developed for this assay yet.

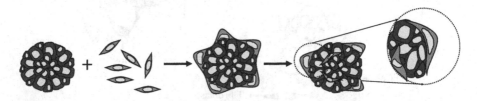

FIGURE 7.6 Schematic representation of spheroid/monodispersed cell invasion assay; cell suspension of one cell type is incubated with spheroids of a dissimilar cell type, and cells in due course invade into the spheroid.

Procedure:

1. Wash tumor cell type A monolayers with PBS, add cell dissociation enzyme (1 mL for T25 and 2 mL for T75 flask), and incubate cells at 37°C for 2–5 min.
2. Check for cell detachment under a microscope and neutralize cell dissociation enzyme with DMEM (5 mL for T25 or 8 mL for T75 flask).
3. Centrifuge cell suspension at $500 \times g$ for 5 min, remove supernatant, and resuspend cell pellet in 1 mL of DMEM to get a single cell suspension without cell clusters.
4. Count cells and dilute the cell suspension to obtain $0.5–2 \times 10^4$ cells/mL.
5. Transfer 200 µL cell suspension to each well in a well plate, incubate at 37°C, 5% CO_2, 95% humidity) for 4–5 days for tumor A spheroid formation.
6. Co-cultivate the formed spheroids with a single cell suspension of a different cell type B.
7. Allow the single cells to attach and wash away unattached cells.
8. Monitor the heterotypic interaction morphologically using fluorescently labeled cells, expressing fluorescent proteins or labeled prior to co-culture with cell tracker dyes. Analyze by (confocal) fluorescence microscopy or fix the cells and perform immunofluorescence or immunohistochemical studies.
9. Quantitatively measure the invasion using cells that can be fluorescently labeled prior to co-culture and analyze by flow cytometry after trypsinization.

Advantages:

• The blockade to be invaded is comprised of firmly organized 3D multicellular aggregates through established tissue-like structures with cell to cell interactions.
• Heterotypic cell interaction.
• The tumor spheroids are highly reproducible in size.
• Closely mimics *in vivo* metastatic condition.
• The invasion assay is performed *in situ* in the same plate as tumor spheroid development, without the need to move them to secondary plates.
• The use of fluorescently labeled cells provides a tool to analyze 3D migration or invasion by (confocal) fluorescence microscopy.
• Enables both high content and high-throughput analyses of tumor cell invasion in combination with the latest technologies of automated image analysis.

Limitations:

• Indirect quantification of invading cells.
• Needs either sample preparation for immunohistochemistry through paraffin embedding and tissue sectioning equipment or deep confocal imaging.
• Demands technical expertise.

- Quantitative detection by flow cytometry requires an early trypsinization phase to take away the outmost cells before single cell dissemination. This is crucial in differentiating the cells attached to the spheroid surface from the actual invading cells.
- Difficult to distinguish attachment from invasion.

7.8 SPHEROID CONFRONTATION ASSAY

Spheroid confrontation assay addresses the interactive and invasive behavior of two dissimilar 3D cell collections [18–20]. Two preformed spheroids prepared from two diverse kinds of cells—one bearing invasive character and the other being non-invasive type, are cultured laterally which finally commence to unify. The infiltration into the opposing spheroid can be either as individual cells or collective or may exhibit a noninvasive phenotype. A distinct margin is observed at the boundary amid the two cell types in the noninvasion case (Figure 7.7). The invading cells can be quantified by either immunohistochemistry analysis or confocal imaging or comprising paraffin embedding and tissue sectioning. The cells can be differentiated immunohistochemically if they express distinct markers. Alternatively, fluorescent labeling of the invading cell category can be done earlier to confrontation and later analyzed either by live imaging or subsequent fixation and additional processing.

Procedure:

1. Obtain monolayer cultures from cell type A and cell type B.
2. Incubate spheroids derived from the two different cell types (A and B) in single round bottom wells.
3. Allow contact of spheroid A and spheroid B on a semisolid nonadhesive agar at 37°C overnight.
4. After adhesion, transfer every individual confronting pair into a 5 mL Erlenmeyer flask filled with 90 mL media.
5. Incubate the culture for 1–7 days depending on cell types on a gyratory shaker at 120 revolutions/min at a relative centrifugal force of $0.25 \times g$ at 37°C under a steady airflow-enriched with 5% CO_2.

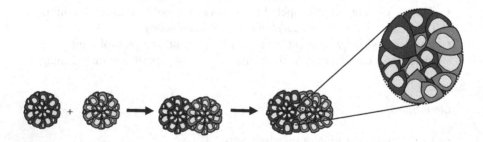

FIGURE 7.7 Schematic representation of spheroid confrontation assay; spheroids made from two dissimilar cell categories are compelled to adhere to each other, later merge, and eventually the invasive cell form infiltrates into the noninvasive cell aggregates.

6. Visualize the invasive reorganization of both the spheroids A and B.
7. Quantify the invading cells either by confocal imaging or immunohisto-chemistry or fluorescent labeling.

Advantages:

* Invasion along tissue-like structures establishing cell to cell interactions.
* Heterotypic cell interaction.
* Reflects closely *in vivo* like carcinoma condition.

Limitations:

* Elaborate postexperimental processing.
* Requires special equipment.
* Formation of multicellular spheroids is a prerequisite.
* Only primary cells and cell lines which are capable of forming spheroids can be used.
* Needs fixation and sectioning or confocal microscopy.
* Time-consuming.

7.9 SPHEROID GEL INVASION ASSAYS

During the implantation of multicellular spheroids into 3D ECM like BME gels or collagen I, the noninvasive cancer cell lines do not display any invasion signs even after 2 long weeks of cell culture and remain as compact spheroids with a discrete margin to the neighboring ECM [21]. Endothelial cells or invasive cell lines present astral extensions from the spheroid and commence to infiltrate into the neighboring matrix (Figure 7.8) [22]. Invasion can be monitored with the aid of live cell imaging and measured by computing the invasive region over a period by using photomicrographs. This can also be achieved by fixing the ECM gels with the invading structures and further processed for immunofluorescence staining and confocal microscopy [23]. Also, the gels can be enzymatically degraded and the cells are

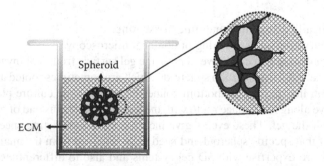

FIGURE 7.8 Schematic representation of spheroid gel invasion assay; invasive cells develop from multicellular spheroidal cell aggregates embedded into the ECM matrix submerged in growth medium to create astral outgrowing structures.

collected for flow cytometry analysis, or protein lysates can be prepared to perform Western blot analysis.

Procedure:

1. Place the well plates containing 4-day-old spheroids on ice.
2. Gently remove 100 μL/well of growth medium from the spheroid plates using a multichannel pipette, avoiding contact with the bottom of the well to minimize disturbance of the spheroids.
3. Gently dispense ECM matrix containing the drug, taking care that the spheroids remain in the center of the well and mix well by swirling gently to avoid the formation of bubbles.
4. Centrifuge the plate at $300 \times g$ for 3 min at 4°C to ensure that spheroids are centrally located in each well.
5. Transfer the plate to an incubator at 37°C and allow the tumor cells to invade from the spheroid body into the ECM matrix.
6. Gently add 100 μL/well of complete growth medium after 1 h.
7. Record the image of tumor invasion or quantify the invasion using flow cytometer.

Advantages:

- Movement of cells along the 3D matrix closely resembles invasion conditions *in vivo.*
- Invasion arises from cell aggregates through established cell to cell associations rather than from individual cells, as observed generally in human cancers.
- Easy detection of exterior boundary of spheroids positioned in the gel using a standard inverted light microscope
- Kinetic measurements of cell invasion possible in live imaging experiments.
- Compatible with molecular biology analysis approaches.

Limitations:

- Technically demanding and time-consuming.
- Needs fixation and sectioning or confocal microscopy.
- Cannot distinguish cell movement on the gel surface from real invasion. For example, cells from certain spheroids which are sometimes rooted at the gel-medium interface or the bottom connected to the tissue culture plates tend to move along the surface route with the least resistance instead of migration through the gel. These events give incorrect results depicting speedy invasion of that specific spheroid and need to be removed from the analysis.
- Needs pre-expertise with 3D gel systems and also to differentiate the surface migration, characterized by cell movement in a single plane and 2D cell morphology, from 3D invasion process.
- Demands high experimental effort.

REFERENCES

1. Y. S. Hwang, K.-K. Park and W.-Y. Chung, Invadopodia formation in oral squamous cell carcinoma: The role of epidermal growth factor receptor signalling, *Archives of Oral Biology*, 57, 335–343, 2012.
2. S. Wang, E. Li, Y. Gao, Y. Wang, Z. Guo, J. He, J. Zhang and Z. Ghao, Study on invadopodia formation for lung carcinoma invasion with a microfluidic 3D culture device, *PloS One*, 8, e56448, 2013.
3. P. Friedl and S. Alexander, Cancer invasion and the microenvironment: Plasticity and reciprocity, *Cell*, 147, 992–1009, 2011.
4. T. Kusama, M. Mukai, M. Tatsuta, Y. Matsumoto, H. Nakamura and M. Inoue, Selective inhibition of cancer cell invasion by a geranylgeranyltransferase-I inhibitor, *Clinical and Experimental Metastasis*, 20, 561–567, 2003.
5. M. Nyström, G. Thomas, M. Stone, I. Mackenzie, I. Hart and J. Marshall, Development of a quantitative method to analyse tumour cell invasion in organotypic culture, *The Journal of Pathology*, 205, 468–475, 2005.
6. V. Härmä, J. Virtanen, R. Mäkelä, A. Happonen, J.-P. Mpindi, M. Knuuttila, K. Pekka, L. Jyrki, K. Olli and N. Matthias, A comprehensive panel of three-dimensional models for studies of prostate cancer growth, invasion and drug responses, *PloS One*, 5, e10431, 2010.
7. M. Zimmermann, C. Box and S. A. Eccles, Two-dimensional vs. three-dimensional *in vitro* tumor migration and invasion assays, Target Identification and Validation in Drug Discovery, *Methods and Protocols*, 227–252, 2013.
8. V. Brekhman and G. Neufeld, A novel asymmetric 3D in-vitro assay for the study of tumor cell invasion, *BMC Cancer*, 9, 1, 2009.
9. A. Albini and R. Benelli, The chemoinvasion assay: A method to assess tumor and endothelial cell invasion and its modulation, *Nature Protocols*, 2, 504 511, 2007.
10. J. Marshall, Transwell® invasion assays, *Cell Migration: Developmental Methods and Protocols*, 769, 97–110, 2011.
11. S. O. Lim, H. Kim and G. Jung, p53 inhibits tumor cell invasion via the degradation of snail protein in hepatocellular carcinoma, *FEBS Letters*, 584, 2231–2236, 2010.
12. V. V. Artym, K. M. Yamada and S. C. Mueller, ECM degradation assays for analyzing local cell invasion, In Sharona Even-Ram and Vira Artym, (eds.) *Extracellular matrix protocols*, 2nd Edn., Humana Press, USA, pp. 211–219, 2009.
13. N. Hamilton, Quantification and its applications in fluorescent microscopy imaging, *Traffic*, 10, 951–961, 2009.
14. I. Ayala, M. Baldassarre, G. Caldieri and R. Buccione, Invadopodia: A guided tour, *European Journal of Cell Biology*, 85, 159–164, 2006.
15. P. Timpson, E. J. Mcghee, Z. Erami, M. Nobis, J. A. Quinn, M. Edward and K. I. Anderson, Organotypic collagen I assay: A malleable platform to assess cell behaviour in a 3-dimensional context, *Journal of Visualized Experiments*, 56, e3089–e3089, 2011.
16. R. C. Inglehart, C. S. Scanlon and N. J. D'Silva, Reviewing and reconsidering invasion assays in head and neck cancer, *Oral Oncology*, 50, 1137–1143, 2014.
17. R. M. Sutherland, J. A. McCredie and W. R. Inch, Growth of multicell spheroids in tissue culture as a model of nodular carcinomas, *Journal of the National Cancer Institute*, 46, 113–120, 1971.
18. K. Hattermann, J. Held-Feindt and R. Mentlein, Spheroid confrontation assay: A simple method to monitor the three-dimensional migration of different cell types *in vitro*, *Annals of Anatomy*, 193, 181–184, 2011.
19. L. de Ridder, M. Cornelissen and D. de Ridder, Autologous spheroid culture: A screening tool for human brain tumour invasion, *Critical Reviews in Oncology/Hematology*, 36, 107–122, 2000.

20. M. Vinci, C. Box and S. A. Eccles, Three-dimensional (3D) tumor spheroid invasion assay, *Journal of Visualized Experiments*, 99, e52686, 2015.
21. H. Dolznig, C. Rupp, C. Puri, C. Haslinger, N. Schweifer, E. Wieser, D. Kerjaschki and P. Garin-Chesa, Modeling colon adenocarcinomas *in vitro*: A 3D co-culture system induces cancer-relevant pathways upon tumor cell and stromal fibroblast interaction, *The American Journal of Pathology*, 179, 487–501, 2011.
22. T. Korff and H. G. Augustin, Tensional forces in fibrillar extracellular matrices control directional capillary sprouting, *Journal of Cell Science*, 112, 3249–3258, 1999.
23. K. Wolf, Y. I. Wu, Y. Liu, J. Geiger, E. Tam, C. Overall, M. S. Stack and P. Friedl, Multi-step pericellular proteolysis controls the transition from individual to collective cancer cell invasion, *Nature Cell Biology*, 9, 893–904, 2007.

8 Anti-Angiogenesis Assays

8.1 INTRODUCTION

Cancer cell proliferation and metastatic spread depend on sufficient supply of oxygen and nutrients and elimination of waste products (Figure 8.1) [1]. Hence, the growth of new and extensive vascular networks is highly significant in metastatic cancer. The sprouting and development of new vasculature from the preexisting blood vessels is termed as angiogenesis, or neovascularization [2]. A tumor cannot grow beyond 2–3 mm without stimulating a vascular supply. New vessels start to develop from the edge of the tumor and then advance into the tumor.

Endothelial cells, which line all blood vessels, are mainly involved in angiogenesis. The angiogenesis process involves several steps; first, degradation of the ECM surrounding the tumor by matrix metalloproteinases (e.g., collagenase), which are up-regulated in some tumor cells allowing the endothelial cells to migrate from their established location by infiltrating through the basement membrane. Once this is attained, the endothelial cells migrate in the direction of an angiogenesis stimulus released from tumor cells and start to proliferate to produce the required number of cells to make a new blood vessel. Subsequently, the fresh extension of these endothelial cells rearranges into a 3D tube-like structure (Figure 8.2) [3]. Each single event that leads to angiogenesis like disruption of basement membrane, migration, and proliferation of cells and formation of tubular structures can be an intervention target and each of these can be studied *in vitro*.

Inhibition of angiogenesis has been shown to suppress tumor growth. Hence, currently the use of anti-angiogenic agents in cancer is under intensive investigation. Most anti-angiogenic agents currently in market being cytostatic rather than cytotoxic stabilize the tumor and avoid tumor metastasis. This feature enables them to be used in combination with cytotoxic drugs, as maintenance therapy in early-stage cancers or as adjuvant management option after definitive radiotherapy or surgery. Suppression of angiogenesis in cancer using appropriate inhibitors can sustain metastases in a state of latency [4–6]. And, surprisingly, to date angiogenesis inhibitors are spared from development of drug resistance.

Quite a few *in vivo* assay models, including the Matrigel plug assay, chick chorioallantoic membrane (CAM) assay, and the corneal angiogenesis assay, have been developed that allow accurate evaluation of the angiogenic response [7,8]. But they suffer from a few disadvantages like inflammatory host responses, time-consuming, and technically demanding. *In vitro* angiogenesis assays are significant in identifying potential angiogenic agents where variables can be controlled and can be selectively focused on endothelial cell migration and proliferation during the angiogenesis process. Endothelial cell proliferation and migration in angiogenesis can be validated using thymidine incorporation and Boyden Chamber Assays, respectively.

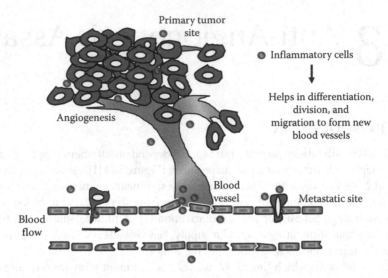

FIGURE 8.1   Schematic diagram depicting angiogenesis and metastatic spread of primary tumor.

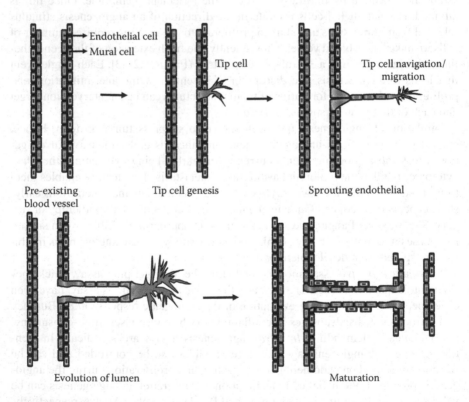

FIGURE 8.2   Schematic representation of angiogenesis.

The angiogenesis assays critically need a more holistic assessment approach, and few *in vitro* assays have been established that allow an improved realistic evaluation of the angiogenesis response.

8.2 ENDOTHELIAL TUBE FORMATION ASSAY

Endothelial tube formation assay is one among the most extensively used *in vitro* assays to model the reorganization stage of angiogenesis and to test angiogenesis inhibitors. The ability of endothelial cells plated at subconfluent densities with the appropriate ECM to form three-dimensional (3D) capillary-like structures (tube formation) mimicking *in vivo* conditions forms the basis of this assay [9]. Culturing endothelial cells on a gel of basement membrane extract (BME) induce their differentiation and tube-like structure formation. Following plating, endothelial cells attach and create mechanical forces on the surrounding ECM support to produce trails that enable cellular migration and eventually result in the formation of hollow tubes or lumens (Figure 8.3). Tube formation is typically quantified by measuring various parameters such as tube number, tube length, tube areas, or branch points in two-dimensional (2D) microscope images of the culture dish. The assay can be employed to investigate the ability of various test compounds to inhibit or promote tube development.

Procedure:

1. Seed 2×10^5 viable endothelial cells (human microvascular endothelial cell lines—IIMEC-1 or HMVEC or primary human umbilical vein endothelial cells—Corning HUVEC-2) in 75 cm^2 tissue culture flask using LSGS (low serum growth supplement) medium 200PRF (medium 200 prepared without phenol red) to a total volume of 15 mL.
2. Change culture medium 24–36 h after seeding, until the culture is ~80% confluent.

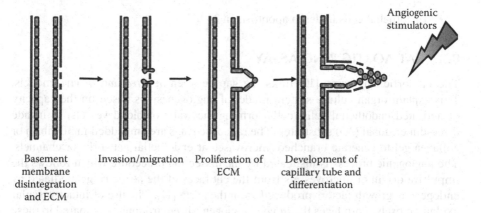

Basement membrane disintegration and ECM   Invasion/migration   Proliferation of ECM   Development of capillary tube and differentiation   Angiogenic stimulators

FIGURE 8.3 Schematic depiction of endothelial tube formation assay; cellular migration leading to lumen development.

3. Trypsinize the cells and transfer the detached cells to a sterile 50 mL coni-cal tube.
4. Centrifuge the cells at $180 \times g$ for 7 min, until the cells are pelleted and remove the supernatant from the tube.
5. Add 4 mL of nonsupplemented medium 200PRF to the cell pellet and resuspend the cells.
6. Plate about $3.5–4.5 \times 10^4$ cells per 200 µL with the test compound and incubate the plate at 37°C in 5% CO_2. (Incubation times may vary.)
7. HUVEC develop well-formed tube networks after 4–6 hours.
8. Stain the cells with a fluorescent dye (calcein AM) and visualize using a fluorescence microscope or with a nonfluorescent dye and visualize directly using a light microscope.

Advantages:

- Moderately easy setup procedure.
- Requires short culture duration.
- Quantifiable results.
- Amenable to high-throughput analysis.

Limitations:

- Selection of cells and matrices are crucial to obtaining consistent and reli-able data owing to significant disparity in tube-forming ability among dif-ferent types of endothelial cells and the support matrices.
- Assay results need to be validated *in vivo* as commercially procured endo-thelial cells have been preselected for their proliferative capacity which does not display any heterospecific cell interactions.
- Use of nonhuman tissues.
- Responses seen with various test compounds may vary with different species.
- Endothelial cells undergo apoptosis in 24 h.

8.3 RAT AORTIC RING ASSAY

The rat aortic ring assay [10] links the gap between *in vitro* and *in vivo* models. This explant organ culture system model of angiogenesis is based on the capacity of activated endothelial cells (cells forming the walls of blood vessels) to invade three-dimensional (3D) substrates. The rat aortic rings are embedded in 3D fibrin or collagen gel to generate branched microvascular endothelial networks or channels. The angiogenic process is instigated by the dissection procedure which follows the impulsive origin of microvessels from the cut faces of the aortic rings mediated by endogenous growth factors produced from the aorta [11]. The use of intact vascular explants closely reproduces the *in vivo* angiogenesis environment compared to those with isolated endothelial cells (Figure 8.4). Microvessels are counted manually. Alternatively, angiogenesis can be quantified using computer-assisted image analysis

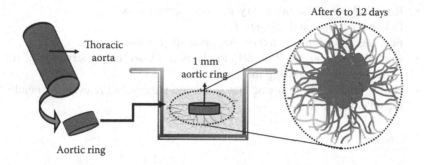

FIGURE 8.4 Schematic representation of rat aortic ring assay; formation of microvessels from aortic ring sections.

[12], which involves the determination of several parameters like the number and dimension (mainly length) of branching micro vessels, the number and spatial distribution of fibroblast-like cells, and the aortic ring size [13,14].

Procedure:

1. Isolate the thoracic aorta from the sacrificed rat by handling only at its ends to avoid trauma to the aorta body.
2. Transfer the aorta into a dish containing serum-free culture medium and remove the surrounding fibro-adipose tissues.
3. Cut away 2 mm sections of aorta both proximally and distally.
4. Cut 1 mm ring segments uniformly and rinse 5–8 times with the culture medium.
5. Remove all blood residues and transfer the rings to fresh cold PBS and maintain on ice throughout the procedure.
6. Add 150 mL cold BME dropwise to each well in a 48-well plate using pre-cooled pipette tips and let the drop solidify at 37°C for around 20–30 min.
7. Keep a single aortic ring on the top center of each dome and incubate for 10 min at 37°C.
8. Add an additional 150 mL BME on top of each ring and incubate further for 20–30 min at 37°C.
9. Add 500 mL human endothelial serum free medium (ECGS – endothelial cell growth supplement) supplemented with 2% fetal calf serum (FCS), 50 units/mL penicillin, 50 µg/mL streptomycin, and test compounds to each well.
10. Incubate the 48-well plate for 12 days at 37°C.
11. Take images of the rings between 6 and 12 days using a standard phase contrast stereoscope.

Advantages:

- Integrates the benefits of both *in vitro* and *in vivo* systems.
- Inhibitors can be tested in a controlled environment.

- Recapitulates all the necessary steps in angiogenesis.
- Relatively simple and inexpensive.
- Provides lot of information on angiogenesis process.
- Resembles an "*ex vivo*" model closely as it allows the preservation of the vessel architecture during the assay.
- Highly superior in terms of biological complexity and relevance to endothelial cultures.

Limitations:

- Angiogenesis *in vivo* starts from microvessels and not from major vessels like aorta.
- Inconsistency in handling of the rings.
- Antigenic responses vary with rings collected from different aortas or with different mice strains and ages.
- Vessel outgrowth is influenced by the amount of residual surrounding tissue on the vessel.
- The 3D angiogenic vessel outgrowth makes it difficult to photograph and quantify.

REFERENCES

1. N. Nishida, H. Yano, T. Nishida, T. Kamura and M. Kojiro, Angiogenesis in cancer, *Vascular Health and Risk Management*, 2, 213–219, 2006.
2. W. Auerbach and R. Auerbach, Angiogenesis inhibition: A review, *Pharmacology & Rherapeutics*, 63, 265–311, 1994.
3. A. M. Goodwin, In vitro assays of angiogenesis for assessment of angiogenic and anti-angiogenic agents, *Microvascular Research*, 74, 172–183, 2007.
4. J. S. Flier, L. H. Underhill and J. Folkman, Clinical applications of research on angiogenesis, *New England Journal of Medicine*, 333, 1757–1763, 1995.
5. T. Boehm, J. Folkman, T. Browderand and M. S. O'Reilly, Antiangiogenic therapy of experimental cancer does not induce acquired drug resistance, *Nature*, 390, 404–407, 1997.
6. S. Benzekry, A. Gandolfi and P. Hahnfeldt, Global dormancy of metastases due to systemic inhibition of angiogenesis, *PloS One*, 9, e84249, 2014.
7. C. A. Staton, M. W. Reed and N. J. Brown, A critical analysis of current *in vitro* and *in vivo* angiogenesis assays, *International Journal of Experimental Pathology*, 90, 195–221, 2009.
8. C. A. Staton, S. M. Stribbling, S. Tazzyman, R. Hughes, N. J. Brown and C. E. Lewis, Current methods for assaying angiogenesis *in vitro* and *in vivo*, *International Journal of Experimental Pathology*, 85, 233–248, 2004.
9. J. A. Madri, B. M. Pratt and A. M. Tucker, Phenotypic modulation of endothelial cells by transforming growth factor-beta depends upon the composition and organization of the extracellular matrix, *The Journal of Cell Biology*, 106, 1375–1384, 1988.
10. R. F. Nicosia and A. Ottinetti, Growth of microvessels in serum-free matrix culture of rat aorta. A quantitative assay of angiogenesis in vitro, *Laboratory Investigation; a Journal of Technical Methods and Pathology*, 63, 115–122, 1990.
11. A. C. Aplin, E. Fogel, P. Zorzi and R. F. Nicosia, The aortic ring model of angiogenesis, *Methods in Enzymology*, 443, 119–136, 2008.

12. K. Bellacen and E. C. Lewis, Aortic ring assay, *Journal of Visual Experiments*, 33, 1564, 2009.
13. S. Blacher, L. Devy, A. Noel and J. Foidart, Quantification of angiogenesis on the rat aortic ring assay, *Image Analysis and Stereology*, 22, 43–48, 2003.
14. S. Blacher, L. Devy, M. F. Burbridge, G. Roland, G. Tucker, A. Noël and J. M. Foidart, Improved quantification of angiogenesis in the rat aortic ring assay, *Angiogenesis*, 4, 133–142, 2001.

9 High-Throughput Screening Assays

9.1 INTRODUCTION

The growing progress in automation technology, data management, combinatorial chemistry, and knowledge on cancer biology drives high-throughput screening (HTS) technology, characterized by selectivity, rapidity, and reliability. It is highly recommended in modern drug innovation processes as it detects more biologically relevant characteristics of active compounds in living systems. The ultimate aim is to gather reliable and quality data in the shortest time possible using the most cost-effective method. Several assays have been developed for screening cell viability and apoptosis with its associated advantages and drawbacks.

HTS assays can be broadly classified into two categories: cell-based assays and target-based biochemical assays [1]. Biochemical assays rely on the specific binding of test compounds to a particular target, but its applications are limited due to difficulty in obtaining highly pure targets and it cannot completely mimic tissue-specific responses. Hence, cell-based assays are used to indicate the toxicity profile of the test compounds in most of all HT screening. NCL60, a panel of 60 human cell lines as recommended by the National Cancer Institute (NCI) that represents nine tissue types, is used for the screening of prospective anticancer molecules. It shares the common characteristics such as cancer specificity, easy to work with, and involves mutations that may affect the experimental results. For phase II screening of test compounds in malignant melanoma and ovary cancer, stem cells with multipotency, self-renewal, and differentiation capacity which are isolated from tumors are available, but their maintenance is difficult compared to normal cell culture.

9.2 CELL CULTURE

Cell-based HTS tests are performed mainly in multiwell plates to escalate the number of wells per plate for HT rates and are handled using an automatic robotic system [2]. HTS platform widely uses static cultures, wherein the intermittent medium change brings in contamination risk and undesirable culture conditions [3]. Hence, modified multiwell plates with integrated microfluidic systems are employed in the high-throughput evaluation of cytotoxicity and long-term effects of anticancer drugs [4,5].

The cell culture modes in HTS include single cells, 2D monolayers, and 3D multilayers or aggregates [6]. Though 2D cell cultures are preferred owing to their low cost and easy management, limited intercellular and cell-matrix interactions can lead to errors in tissue-specific responses, whereas 3D cultures better recapitulate *in vivo* conditions in intercellular and cell-matrix interactions allowing transport of nutrients and metabolites, cell migration, invasion, morphogenesis, an increase in drug resistance, and so on [7]. They provide more useful information during the anticancer drug screening and development studies like side effects of a drug in neighboring cells [8].

9.3   HIGH-THROUGHPUT CYTOTOXICITY AND APOPTOSIS SCREENING IN 2D CULTURES

Cytotoxicity measurements based on quantification of viable cell number using conventional hemocytometer are time-consuming and labor intensive. Flow cytometry and invasive chemicals like Trypan blue and neutral red have comparatively low throughput. As HTS involves the use of a small amount of culture medium, detection of cytotoxicity is achieved by two noninvasive, online monitored methods: electrochemical and optical.

The electrochemical method employs biosensors with integrated biological recognition elements and electrochemical transduction units which are mainly two types: (1) that detects electron generation, charge transfer, and changes in ionic concentration caused by redox reactions and changes in metabolic products like lactic acid, CO_2, O_2 glucose, and so on in viable cells [9]; and (2) electrochemical impedance spectroscopic techniques which can monitor cell number, viability, morphology, apoptosis, and cell adhesion. However, this method cannot provide information on cytotoxic mechanisms of test compounds and is not amenable to 3D cell cultures.

Optical detection using colorimetric methods is invasive, time-consuming, less sensitive, laborious, demands several scheduled additions of chemicals that can disturb the targeted cells, and can only provide endpoint data. Hence, fluorescent methods that can be miniaturized and have higher sensitivity compared to luminescent methods are employed for large-scale cell-based HTS measurements of cell activities and toxicity [10]. Reporter gene methods employing GFP (green fluorescent protein) allow real-time, automated, noninvasive detection and quantification of cell proliferation as well as specific cellular functions [6]. Moreover, GFP coupled with dsRed (Disco soma species red) fluorescent protein are employed for multiplex or two-color assays [11]. In addition, noninvasive assays using cDNA encoding a fluorescent protein offer HT analysis of cell proliferation and death kinetics. Fluorescence resonance energy transfer (FRET) enables studying the activation of caspase-3 or apoptosis in viable cells [12] by using two fluorescent proteins attached to a peptide linker comprising a caspase-3 cleavage site. It is based on alterations in the emission wavelength owing to the transfer of energy between the two close fluorophores. Commercial whole-cell autofluorescence-based HTS systems [13,14] are simple, fast, and use laser scanning imaging techniques combined with both fluorescence microscopy and quantitative image analysis to examine the living cells and quantify intracellular proteins [15]. However, their use is limited due to high costs, relatively low capacity, and nonsuitability for 3D cell cultures [16].

9.4 3D CELL-BASED HTS ASSAYS

Changes in the conventional 96-well culture environment like pH and other auto-fluorescent constituents that exist in the culture medium generate weak and fluctuating fluorescence signals [17] and efficiently mask viable cell GFP signals rendering it unpredictable for evaluating cell proliferation or cytotoxicity. Hence, GFP-expressing cells are cultured in a PET support in a modified well, which considerably escalates the total cell number/unit area [18], and such a 3D culture can afford a 20-fold greater cellular fluorescence. This considerably improves the signal-to-noise ratio as the cells are confined in the scaffold at the interior of the well and the background fluorescence can be independently measured and subtracted to provide the actual viable cell signal. In addition, the 3D fluorescent cell culture platform offers extremely reproducible growth kinetic results, which can be more consistently employed to study cytotoxicity effects of chemicals on the proliferation of cancer cells.

Microfluidic micro-bioreactor array systems operated with continuous perfusion are integrated with modern developments in tissue engineering, microfluidics, and microfabrication. This can be employed to culture carcinoma and stem cells in 3D microfabricated supports for extended study of anticancer drugs in a 3D setting resembling an *in vivo* environment. 3D cell-based microfluidic HTS assays facilitate the use of fluorescent cells, provision for on-chip serial dilutions and mixtures of multiple drugs to be tested concurrently on a single chip and feasibility to culture various types of cells in different but interconnected compartments to appraise intercellular and cell extracellular environment interactions on a microfluidic chip [19]. These benefits provide drug responses in a biosystem level that was only gathered by *in vivo* tests to date. Microfluidic systems are furnished with entirely automated external physical controls and online detection devices that can provide enhanced data value and reduced assay duration as well as the cost for high-throughput screening assays.

REFERENCES

1. W. F. An and N. Tolliday, Cell-based assays for high throughput screening, *Molecular Biotechnology*, 45, 180–186, 2010.
2. S. A. Sundberg, High-throughput and ultra-high-throughput screening: Solution- and cell-based approaches, *Current Opinion in Biotechnology*, 11, 47–53, 2000.
3. M. H. Wu, S. B. Huang and G. B. Lee, Microfluidic cell culture systems for drug research, *Lab Chip*, 10, 939–956, 2010.
4. V. Lob, T. Geisler, M. Brischwein, R. Uhl and B. Wolf, Automated live cell screening system based on a 24-well-microplate with integrated microfluidics, *Medical and Biological Engineering and Computing*, 45, 1023–1028, 2007.
5. S. Y. C. Chen, P. J. Hung and P. J. Lee, Microfluidic array for three dimensional perfusion culture of human mammary epithelial cells, *Biomedical Microdevices*, 13, 753–758, 2011.
6. S. T. Yang, X. Zhang and Y. Wen, Microbioreactors for highthroughput cytotoxicity assays, *Current Opinion in Drug Discovery and Development*, 11, 111–127, 2008.
7. S. Basu and S. T. Yang, Astrocyte growth and glial cell line-derived neurotrophic factor secretion in three-dimensional polyethylene terephthalate fibrous matrices, *Tissue Engineering*, 11, 940–952, 2005.

8. L. A. Gurski, 3D matrices for anti-cancer drug testing and development, *Oncology*, 25, 20–25, 2010.

9. W. Nonner and B. Eisenberg, Electrodiffusion in ionic channels of biological membranes, *Journal of Molecular Liquids*, 87, 149–162, 2000.

10. P. Gribbon and A. Sewing, Fluorescence readouts in HTS: No gain without pain? *Drug Discovery Today*, 8, 1035–1043, 2003.

11. M. Wolff, J. Wiedenmann, G. U. Nienhaus, M. Valler and R. Heilker, Novel fluorescent proteins for high-content screening, *Drug Discovery Today*, 11, 1054–1060, 2006.

12. N. P. Mahajan, D. C. Harrison-Shostak, J. Michaux and B. Herman, Novel mutant green fluorescent protein protease substrates reveal the activation of specific caspases during apoptosis, *Chemical Biology*, 6(6), 401–409, 1999.

13. Cellomics Inc, Pittsburg, PA.

14. BD Biosciences, San Jose, CA.

15. V. C. Abraham, D. L. Taylor and J. R. Haskins, High content screening applied to large-scale cell biology, *Trends in Biotechnology*, 22(1), 15–22, 2004.

16. S. A. Haney, P. LaPan, J. Pan and J. Zhang, High-content screening moves to the front of the line, *Drug Discovery Today*, 11, 889–894, 2006.

17. P. Girard, M. Jordan, M. Tsao and F. M. Wurm, Small-scale bioreactor system for process development and optimization, *Biochemical Engineering Journal*, 7(2), 117–119, 2001.

18. X. Zhang and S. T. Yang, High-throughput 3-D cell-based proliferation and cytotoxicity assays for drug screening and bioprocess development, *Journal of Biotechnology*, 151, 186–193, 2011.

19. R. Zang, D. Li, T. I-Ching, J. Wang and S. Yang, Cell-based assays in high-throughput screening for drug discovery, *International Journal of Biotechnology for Wellness Industries*, 1, 31–51, 2012.

Index

Printed in the United States
by Baker & Taylor Publisher Services